ENVIRONMENTAL SCIENCE

ECOSYSTEMS

GLOBE FEARON
EDUCATIONAL PUBLISHER
PARAMUS, NEW JERSEY

Paramount Publishing

CONTRIBUTING WRITERS
Mary Jo Diem
Janice Haymes
Lee Hunter
Seth Shulman

REVIEWERS
Stanley Blais
Robert Hubert
Jeff Manker
Lynn Young

CONSULTANTS
Maria Arguello
Alan Ascher
Janet Lyons-Fairbanks
Patricia Marinac
Patricia Neidhardt

Cover Design: Richard Puder Design
Cover Photo: Image Finders/Dave Watters
Electronic Technical Art: Siren Design
Executive Editor: Joan Carrafiello
Project Editor: Doug Falk
Production Manager: Penny Gibson
Manufacturing Supervisor: Della Smith
Senior Production Editor: Linda Greenberg
Production Editor: Alan Dalgleish
Electronic Interior Design and Production: Margarita Giammanco
Art Direction: Nancy Sharkey
Photo Researcher: Jenifer Hixson
Marketing: Sandra Hutchison

Photo acknowledgments appear on page 92.

Printed in the United States of America
4 5 6 7 8 9 10 99 98

ISBN 0-835-90553-5

GLOBE FEARON
EDUCATIONAL PUBLISHER
PARAMUS, NEW JERSEY

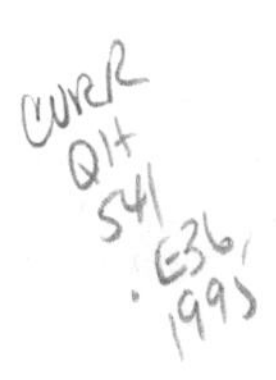

TABLE OF CONTENTS

ECOSYSTEMS

Case Study

The Ecologists from Cass Tech

Have you ever wondered what your neighborhood would look like if people had never lived there? What kinds of trees would grow there? What kinds of animals would live where your home is?

Until recently, Amber Strickland knew little about the trees and animals of her home town. Born and raised in the heart of Detroit, Amber had almost never been outside a city environment. "I had never paid much attention to nature," she said. "Nature wasn't something important to me. It didn't really seem to affect me."

Amber's views changed when she began to learn about and work at Belle Isle. This is a large island, just two miles east of downtown Detroit, surrounded by the Detroit River. The entire 1,000-acre island is a recreation and natural area. At least half of it, run by the nature center on the island, is almost untouched by humans. Of the wild parts of Belle Isle, more than 300 acres are wetland and forest. This is where people can find tall white oak and silver maple trees, plus turtles, snakes, hawks, owls, and many other kinds of wildlife.

Amber still clearly remembers her first visit to the "wild side" of Belle Isle on a clear, sunny day in May. She was just several miles from her home in the city, but Amber thought she might as well have been visiting another planet. She got her first look at the habitat of birds and turtles. "On the nature trails, when you get close to plants and animals in the wild," she says, "it is like a whole different world. And seeing that world has changed my attitudes a lot."

Amber's work on Belle Isle is part of an ambitious program in environmental science at Cass Technical High School in Detroit. Students in the Urban Environmental Education Program visit Belle Isle frequently in the spring. They learn about the ecology of Belle Isle's plants and animals.

Some of the high-school students—like Amber—work as assistants at the Belle Isle nature center during the summer months. They prepare nature activities to use when they guide visitors on walks. They help the staff of the nature center with routine tasks. They also work together on special ecology projects.

Saving Turtles

Amber participated in a project to help restore Belle Isle's turtle population. Turtles lay their eggs in the shallow sands and soils of the island. Several kinds of predators, such as raccoons, eat turtle eggs when they uncover them. In the turtle project, Amber and others collected turtle eggs from around the island. They kept the eggs in captivity until they hatched, and then released the baby turtles in their natural habitat.

Students did other work on the ecology of the island. Amber, for example, sampled water for pollutants and then shared the data with students in other areas—even other countries—using computers linked together over telephone lines. She learned to identify and classify many different plant and animal communities. She visited a nearby field testing station of the Environmental Protection Agency and met with scientists there.

During fall and winter, Amber and her fellow students took an urban science course at Cass Tech. As a part of the program, the high-school students taught elementary-school pupils about environmental science. Serving as teaching assistants in nine separate Detroit elementary schools, Amber and other students help to inspire classes of 4th and 5th graders about the environment. As Amber's teacher Randall Raymond puts it: "I think students often know better than teachers that science education shouldn't be taught as something separated from their daily lives. Right from the start, students should be able to use what they learn to help others and themselves to live in better harmony with their environment."

For students from an urban area like Detroit, Raymond says, Belle Isle plays an important part in this effort. "Students need more than a 'one-time trip to the zoo' approach to nature studies. They need to come to a place like Belle Isle in all different seasons and really get a feel for it." In the fall, he says, with the foliage and falling leaves, students can best learn to identify and classify different species of trees. In the spring, there are wild flowers and aquatic systems to explore.

Owl Pellets

Winter is the best time to investigate birds of prey and the foods that allow them to survive. In addition to birds such as red-tailed hawks, Belle Isle is home to great horned owls. Like other birds of prey, owls regurgitate the undigested bones and fur of small animals they eat. The undigested body parts are spit out in compacted balls called pellets.

Amber and her classmates gathered the owl pellets from the grounds of the nature center. They investigated the pellets' contents to identify first-hand the foods that owls eat.

For Amber, working in a natural ecosystem was an exciting new adventure. She began to see Detroit in a new way. It was the first time that she really thought about the city as something built by people. "I'm seeing the same things in the city I always saw, but I never paid attention to them before. My experience at Belle Isle has helped me to make a lot of connections. I have seen how so many of the things we have done have hurt the environment—the water supply, the soil. Now almost everything I do I think: how is it affecting the environment. I think about whether something is biodegradable or recyclable. I don't use Styrofoam. I don't litter anymore. I try to ride the bus instead of always driving in someone's car. "

Regina Wertz has participated in the program with Amber and speaks equally strongly about the way in which Belle Isle and the Urban Environmental Education Program have given her a new lens through which to look at the city. She has helped set up a paper recycling program in her computer class and a school-wide recycling program for people's returnable bottles and cans.

"I think the most important thing about learning about the environment is learning to see how everything is related," Regina says. She and Amber were some of only a handful of students to attend a state conference of science and environmental educators. "One of the most exciting things we did," Regina says, "is we worked in groups on a problem about land use." The people at the conference debated the pros and cons of building a mall in a particular area compared with building a chemical plant. "We were asked to think about the impact of each on the environment," Regina says. At first, she remembers, people focused on the pollution that the plant would cause. Then, she and her group explored the impact of the mall. It would pave over a huge area and cause a great deal of air pollution from car traffic.

"We broke down the problem piece by piece, deciding what would be really costly to the environment—in the soil, for instance, what particular pollutants would be involved in each case. It was really interesting. Like so many parts of this program," she says, "it helped me think about the big picture."

1 WHAT ARE ECOSYSTEMS?

Ecosystems are everywhere, including cities. This park environment includes water, soil, and living organisms on land and in the water.

1.1 Environment and Ecosystems

It seems that everyone is talking about the environment. Television programs and newspapers often run stories on the environment. Parents are concerned about pollution and their children's future environment. Politicians debate how best to protect the environment. You are preparing to learn about the environment. But what exactly does *environment* mean?

Your environment consists of your surroundings. Your environment can be the building that you are in, whether it is your home or school. It is the neighborhood, and the city or town in which you live, too. But it is also much more than that. Your environment includes the things that support life, such as air, soil, water, and the chemicals in them.

Environment includes physical conditions such as climate, elevation, and other geographical features of your world. Other living things—the plants, animals, and microbes—are also a part of it. The energy that keeps everything going is a part of your environment, too.

Although most of us live in an environment made by humans, that environment is supported by the natural world. We depend on healthy soils to produce the food we eat. That means that soil must be cared for. Soil that is poisoned and worn out can't grow food.

Lakes, streams, and rivers must be protected to provide people with a clean, safe drinking-water supply. This requires not only protecting these bodies of water, but also the land around them.

A garden in a yard is an example of an ecosystem controlled by people.

We all need clean air to breathe. This means that we need to understand how our activities affect the quality of our air. The burning of fuels for making electricity and for transportation contributes to air pollution. So it is important that we understand how our environment works. We want to be sure that we will continue to survive on a healthy planet.

WHAT ARE ECOSYSTEMS?

Scientists divide our environment into units called ecosystems. An ecosystem can best be defined as a group of plants, animals, and **microbes** that interact with one another and the physical environment.

Ecosystems can be small or large. The entire earth is an ecosystem called the biosphere. A pond is an ecosystem; so is a river. These are called aquatic ecosystems. Other aquatic ecosystems are oceans, streams, and marshes. Land ecosystems are called terrestrial ecosystems.

While all of these ecosystems look very different from one another, they have some features in common. The living things in these places interact with each other and with the environment in a way that sustains the ecosystem.

Human actions shape many ecosystems. Your backyard and schoolyard are ecosystems. A farm is a carefully controlled ecosystem. In the case study, you read about students studying an island ecosystem within a city.

City neighborhoods and parks are ecosystems. Humans, plants, and animals are all part of the ecosystem on Belle Island in Detroit, Michigan. In urban ecosystems, the factors that support life are much affected by the activities of humans.

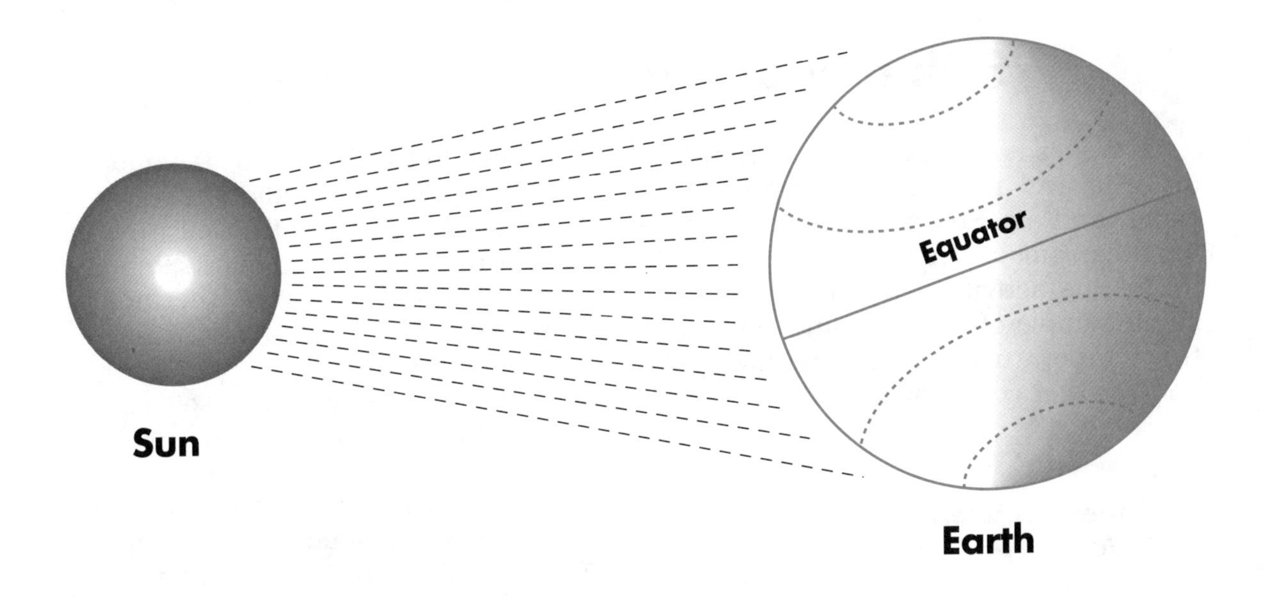

Sunlight is an essential abiotic part of an ecosystem. The amount of sunlight that reaches the earth's surface is influenced by latitude.

1.2 What Makes Up an Ecosystem?

Ecosystems can be divided into the components or parts that make them up. Scientists study the parts in order to better understand ecosystems. **Abiotic** components are the nonliving parts of ecosystems. They are the chemical and physical conditions that make up the environment. These conditions include rainfall, temperature, light, elevation, depth, rocks and minerals, soil, and chemical **nutrients**, and the slope of the land.

ABIOTIC PARTS OF ECOSYSTEMS

Moisture Water is an abiotic component. All plants need water. Trees need a lot of water. Where there is not enough rain to support the growth of trees, grasses grow instead. In a dry area, trees may grow only along river banks. Forests are thick with trees and other plants because they receive so much rainfall. The amount of water in an ecosystem, then, affects where trees grow. It also affects how many trees grow.

Temperature Heat and cold are important abiotic factors, too. In places with cold winters, water is in the form of ice or snow much of the time. Water as ice or snow is not available to plant roots, unless it melts. If water and the surface of the soil are frozen, then plants cannot grow.

Latitude Geography affects abiotic factors such as temperature and rain. Latitude is the location on the earth moving from north to south. The

Soil: a vital part of ecosystems.

The soil is a mixture of minerals, organic and inorganic compounds, water, and air. Soil contains many living organisms as well.

In this activity you will examine a small sample of soil and find out how the temperature changes with depth and weather conditions.

Materials

5 thermometers
1 dowel rod
1 ruler
1 watch
1 ball of string
4 stakes

Procedure

1. Select a plot of land around your home or school. Make sure the land is not in a high-traffic area.
2. Make a map of the area immediately surrounding your plot. List and describe the abiotic and biotic factors that affect your plot.
3. Use the stakes and string to mark off a square foot of the land.
4. Use the ruler to mark off five 3-cm increments on the dowel rod as shown in the figure. The 0 cm mark should be at one end of the rod.
5. Press the rod into the soil in your plot up to the 3 cm mark. Remove the rod and insert the thermometer into this hole. This thermometer is 3 cm deep.
6. Press the rod into the soil again, this time press it into the soil up to the 6 cm mark. Insert the second thermometer into this hole. This thermometer is 6 cm deep.
7. Repeat Step 5 until you have inserted three more thermometers at depths of 9 cm, 12 cm, and 15 cm, respectively.
8. Record the date and time on the chart provided by your teacher. Observe and record the weather conditions at the time of each reading. Include air temperature, sky conditions, and any precipitation that may be occuring.
9. Then, read each thermometer and record the temperature in the appropriate place on the chart.
10. Repeat Steps 8 and 9 every hour for at least four hours.

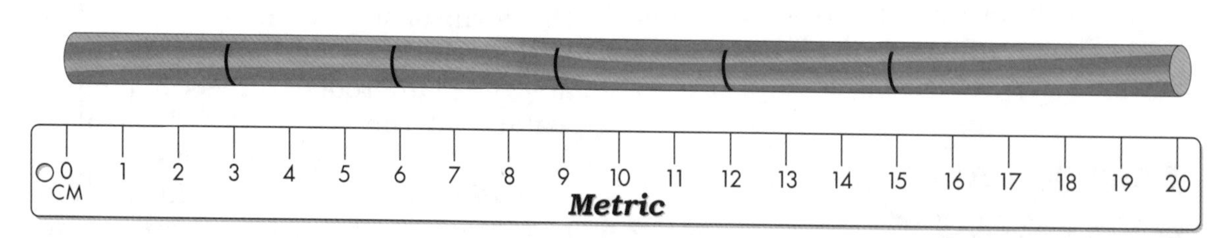

Use a rod or dowel to make holes in the soil for measuring temperature at different depths.

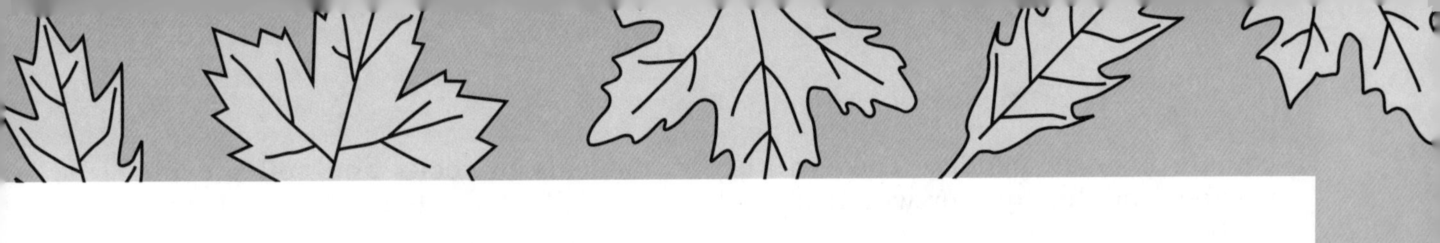

DATE	TIME	WEATHER CONDITIONS	SOIL TEMPERATURE (°C)				
			3 cm	6 cm	9 cm	12 cm	15 cm

Conclusions

1. Graph the temperature of the soil versus the time of the reading. Use a different color line on your graph for each depth that the readings were taken.
2. How is the temperature of the soil affected by changes in depth? How is the temperature of the soil affected by the time of day?
3. How are the temperature readings affected by weather conditions?

For Discussion

1. Compare the maps and results of your classmates. How do their results compare to yours? How can you explain any differences?
2. Do you think temperature variation would affect the biotic parts of the soil ecosystem? Why or why not?

Extension

Predict what would happen if you repeated this experiment at various times throughout the year. Then do the experiment to find out if your prediction is correct.

equator is in the middle. The earth is warmest at the equator which has a latitude of 0. Temperatures become cooler as you move north or south from the equator to the poles.

Length of Day The amount of light that is available to plants is also determined by latitude. Plants at the equator have the same length of day all year. At higher latitudes, there is less light and shorter days during winter. This creates a shorter growing season. All plants need light to make their food. Without enough light, plants cannot grow. Fir trees in the northern forests can survive not only the colder temperatures, but also a shorter growing season.

Aquatic ecosystems are different depending on latitude and the depth of the water. In the oceans, different kinds of creatures are found at different depths because there are differences in temperature and light. Light only penetrates the water to a certain depth. Below that depth, no plants can grow. Even in a pond or lake, plants at the shallow edges are different from plants in deeper parts.

Gases Dissolved oxygen and carbon dioxide are important chemicals in aquatic ecosystems. The availability of these chemicals helps to determine the types of plants and fish found in a river, stream, lake, or pond. Certain kinds of pollution reduce the amount of dissolved oxygen in water. Fast-swimming fish, such as trout or bass, require water that contains a lot of oxygen. Bottom-dwelling fish, like carp, move slowly and are often found in polluted streams.

Minerals The characteristics of the rocks and soil influence the kind of ecosystem found in a particular place. Chemicals in rocks and soil provide minerals that plants need. This is true in aquatic ecosystems, too. Chemicals in the soil on the bottom provide nutrients for plants. Erosion from the land washes nutrients from rocks and soil into a body of water. These nutrients are used by plants and then by the animals that eat them. The slope of the land determines how much erosion takes place.

Soil particles come from eroded rock. Different kinds of rocks erode to produce different kinds of soils. Their mineral makeup will be different and so will the particle size. The size of soil particles affects how well the soil holds water. Some soils hold water well. Sandy soils do not. They drain very quickly, even after a heavy rain.

Many of the soils found in Florida are sandy because most of Florida used to be under the ocean. As the ocean receded, it left sand dunes. The tops of the ancient sand dunes are called ridges. The ridge ecosystems are high and dry. It rains a lot in Florida, but ridge soils drain very quickly.

The plants in the ridge ecosystems have a lot of the characteristics of desert plants because they have to be able to conserve water. There is a lot of evaporation with the heat of the sun, even on a humid day. Animals that live in the soil of this ecosystem, like the gopher tortoise, dig deep burrows. This is one way to cope with the heat and dryness.

The abiotic factors in an ecosystem are closely connected to each other. Scientists can learn a lot about how living things interact with the abiotic parts of their environment by studying them in the laboratory.

A tree is an example of an autotroph. **A whale is an example of a heterotroph.**

In natural ecosystems, the effects of the many abiotic factors in an ecosystem are often hard to separate. The Field Study in this Chapter demonstrates some of the ways that different abiotic factors in an ecosystem can affect each other. In this activity, you will look for some relationships between soil temperature, soil depth, and weather conditions. Ecosystems are affected by the combination of abiotic factors found there.

BIOTIC FACTORS IN ECOSYSTEMS

The living organisms in an ecosystem are called **biotic** factors. Some organisms are familiar plants and animals, living things you have seen in parks, museums, zoos, and botanic gardens. Many organisms in ecosystems are not familiar. For example, they may be tiny microbes that you can only see with a microscope. Unfamiliar or not, the plants, animals, microbes, and other kinds of living organisms are not the same in all ecosystems. Just as abiotic factors differ from place to place, so do biotic factors.

Biologists divide these biotic factors according to their activities in ecosystems. In general there are two groups of organisms that have different functions. One group is called the **autotrophs**. The word autotroph means "self-feeder." This word is used to describe organisms, such as green plants, that make their own food. Autotrophs use energy from the sun to do this. Some autotrophs use energy from other sources in the environment to make food.

The second group of living organisms is called the **heterotrophs**. The word *heterotroph* means "other feeder." This term describes an organism that cannot use energy to make its own food. Instead, it must eat other living things to get food. Animals are heterotrophs. So are a group of microbes called bacteria.

Lab Study

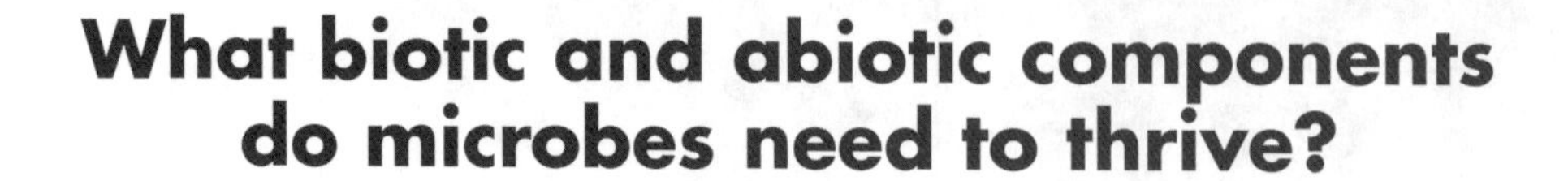

What biotic and abiotic components do microbes need to thrive?

In this activity, you'll find out what microbes need to grow in a mini-ecosystem.

Materials

8 plastic petri dishes with lids
1 medicine dropper
1 role tape
2 slices of bread
50 mL of milk
1 green vegetable, cooked
1 slice of lunch meat
1 piece of fruit (apple or orange)
1 hand lens

Procedure

Part 1

1. Place a square piece of bread (about 3 cm square) in each of four petri dishes. Label the dishes A, B, C, or D.
2. Use the medicine dropper to place 5 or 6 drops of water on the pieces of bread in dishes A and C. The bread should be moist, not soaked.
3. Put the covers on each of the petri dishes. Tape the dishes closed.
4. Place the petri dishes marked C and D in a refrigerator. Place the petri dishes marked A and B in a warm place out of direct sunlight.
5. Table 1 summarizes the environmental conditions for each of the four petri dishes.

TABLE 1
A - bread, warm, water
B - bread, warm, no water
C - bread, cool, water
D - bread, cool, no water

Part 2

1. Label and prepare the following four petri dishes:
   E - vegetable, water, warm
   F - meat, water, warm
   G - milk, water, warm
   H - fruit, water, warm
2. Cover each petri dish and tape it closed.
3. Place these containers in a warm place out of direct sunlight.

Part 3

1. Check all eight containers every day for about 2 weeks. Look for changes that are taking place in your food samples.
2. Record the changes that you observe in the chart provided by your teacher.

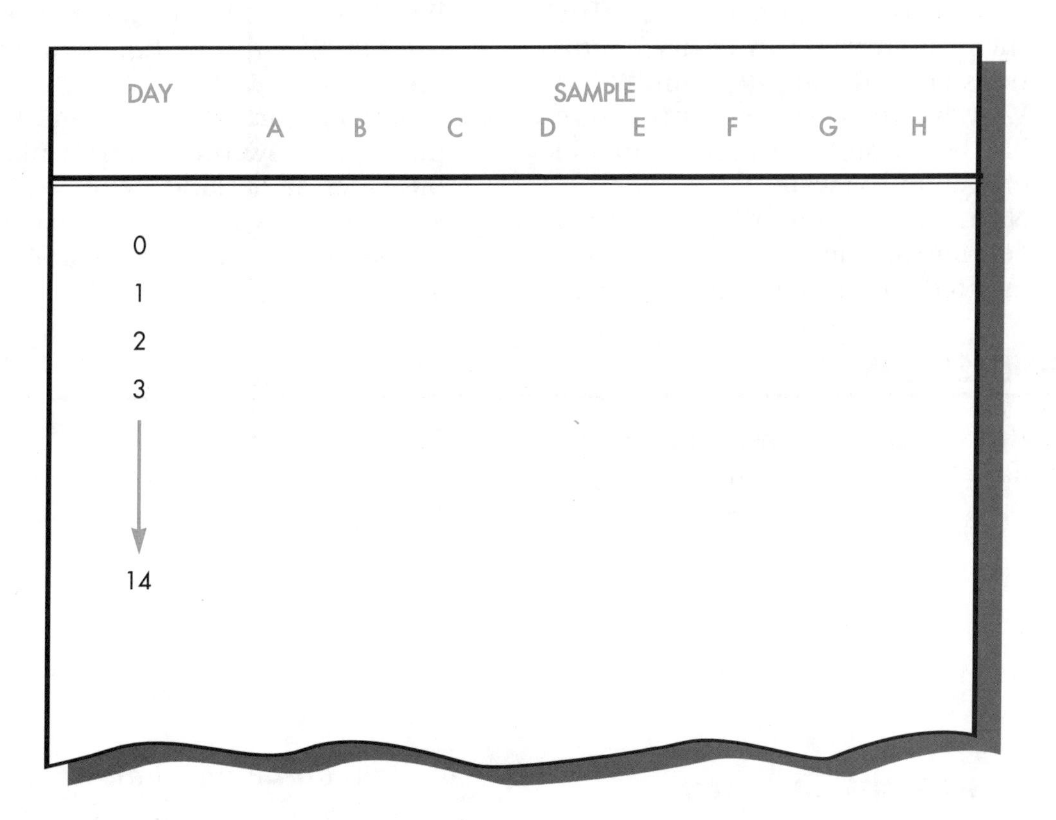

DAY	SAMPLE A	B	C	D	E	F	G	H
0								
1								
2								
3								
↓								
14								

Conclusions

1. Under what conditions did the microbes grow best?
2. Were there any conditions where microbes did not grow at all?
3. What are the ideal biotic and abiotic conditions of the microbes' ecosystem?

For Discussion

1. What is the role of microbes in an ecosystem?
2. Microbes are both harmful and helpful. Explain why.
3. What is the origin of the microbes that grew in your petri dishes?

Extension

Research current and historical ways to preserve foods. Explain the effectiveness of these methods in terms of the ideal environment for growing microbes.

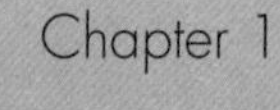

SECTION REVIEW

1. What kinds of adaptations would you expect to see in the bodies of sand-dune plants that enable them to survive in the dry conditions of sand, salt, and wind?
2. Why would fast-swimming fish like bass and trout require a lot of oxygen in the water?
3. Natural ecosystems are usually self-sustaining. Is an ecosystem created in the laboratory a self-sustaining ecosystem? Why or why not?
4. What is the relationship between the equations for photosynthesis and respiration?
5. Where on earth do you think that most photosynthesis takes place: on water or on land? Why? What kinds of organisms are responsible for this photosynthesis? Be specific.

FOR DISCUSSION

What would happen in an ecosystem if there were no decomposers?

1.3 Human Culture Affects Ecosystems

Modern people have a potential that has never before existed for changing the earth's environment. All living organisms have an impact on the environment. Microbes that cause disease, for example, can wipe out whole populations.

Human beings, however, have a much greater ability to change their environment than other species have. There are a number of reasons for this. One reason is that humans walk upright, on two legs instead of four. This frees our hands to do all of the things that we do.

How do human hands affect ecosystems?

TOOLS CHANGE ECOSYSTEMS

Our skillful hands have so much power to change our world. Think about all of the things you do with your hands. Right now, you are using them to hold this book. The structure of our hands allows us to manipulate materials and to make and use tools. With tools, humans can manipulate the environment.

Many animals have hands that work the way that ours do. Monkeys, apes, and other primates have hands like ours, too. The difference between humans and these other animals is in the human brain.

The human brain is very complex. Scientists are still trying to learn about the origins and development of the human brain. The largest part of our brain is the cerebral cortex. This is the front and top portion of the brain. Language, thought, ideas, and plans all start in the cerebral cortex. This part of our brain has allowed us to develop our **culture** and **technology**. We can work together, make tools, make clothing, produce art, plan a hunt to find food, and help each other raise our children. Our brain has enabled us to manipulate our world in ways that no other species can.

Another reason for human success in surviving on the planet is the use of fire. Once people learned to make fire, they were able to move into harsher climates and cook foods. Cooking removes poisonous chemicals from plants. Cooking food increased variety in the diet and allowed humans to survive in new habitats.

Fire also made it possible to use metals to make useful objects. It also brought people together into groups and communities, telling stories and sharing knowledge around the fire at night.

From the moment humans appeared on the planet, they began having an impact on their environment. At first, the impact was small. Early humans lived in small groups that were nomadic. Nomads are groups of people that move from place to place. Usually, they move in order to find food. These groups of people were hunter-gatherers. They did not **domesticate** plants or animals for food. They would go out and collect their food from nature.

As their technology improved, hunter-gatherers began to have more of an impact on their environment. Groups of hunter-gatherers became larger. Hunters were more skillful because they could work together. Language enabled them to do this. Hunters could quickly wipe out game animals in an area. Some scientists believe that hunter-gatherers in North America helped bring on the extinction of a lot of large mammals after the Ice Age. These hunters may have helped to cause the extinction of animals like the woolly mammoth and the mastodon.

Some scientists believe that in later hunter-gatherer societies, the population began to grow. This meant that in order to find enough food, people had to travel farther. Perhaps, to solve this problem, about 10,000 to 12,000 years ago, human cultures began domesticating plants and animals and growing them for food. This activity is called agriculture, or farming.

FARMING CHANGES ECOSYSTEMS

Farming began slowly. Work was done by hand with human muscle power. People no longer had to wander to gather plants for food. They could grow them near home.

The domestication of animals was an important change, too. Animals such as pigs, chickens, and goats were domesticated to provide meat and milk. Humans began making changes in other species. Humans would breed animals, choosing traits in the animals that were most desirable for human needs. The domestication of animals helped some nomadic groups because animals could be used for work and for food.

The grazing of cattle, sheep, and goats sometimes had a serious effect on the environment. Shepherds often would allow their animals to overgraze the land. Domestic animals such as cattle and sheep do not graze the way deer and other wild animals do. Domestic animals do more damage to the grasses and other vegetation that they eat. Sometimes, the plants do not grow back. Bare soil erodes, damaging that important abiotic part of the ecosystem. Overgrazing, especially on fragile, dry lands, is still a big environmental problem in the world today.

Shepherds changed their ecosystems in another way. They often burned and cleared forests to make more grassland for their animals. Forests are important because they capture rainwater so that it soaks slowly into the soil. This prevents soil erosion. Forests also take in the water and release it into the atmosphere, where it becomes rain again.

Forests protect soil and water resources.

Some farmers used fire to clear their land for planting. This type of agriculture is called slash-and-burn agriculture. It is a method of farming still used in many tropical parts of the world. Farmers burn a section of the tropical forest to clear it for planting crops.

Tropical soils are poor in nutrients. Most of the nutrients are in the trees and other plants. Burning the plants makes ash that is rich in nutrients. This is added to the soil for the crops. The land, though, can only be farmed for a few years before the soil loses its nutrients. Heavy rains wash the nutrients out of the soil, and the land bakes in the sun and becomes more like rock than soil. Then these farmers must move to another part of the forest and clear it to grow their food. This gives the forest time to grow back.

This kind of farming works very well if there are not a lot of people clearing the forest. In many tropical areas now, there are too many people practicing slash-and-burn agriculture. This is one of the major threats to tropical rain forests today.

Agriculture changed as human technology and culture changed. People domesticated animals such as horses, oxen, and donkeys. These animals helped with the work on the farm. Now people could plant larger areas of land and provide food not only for their own families but for their neighbors, too.

People have cut forests to clear land for growing crops. This has helped human populations to grow, and has damaged ecosystems, too.

Nomads had few possessions because they had to carry all of their belongings with them as they moved around to find food. Farming allowed people to settle in a permanent place. Cities and towns grew out of farming communities. Once farmers could feed more people than just their own families, some people could do jobs other than farming. The division of labor in communities began. Some people farmed to produce food for the community. Others practiced trades such as carpentry, baking, blacksmithing, or making clothing. People began to acquire more material possessions because new objects were being made through the crafts. People could own more material possessions because they were living in one place.

Agriculture helped the human population on earth begin to grow much faster. Children were needed in agricultural societies to work on the farms, so the population grew. It became important for children to inherit the land from their parents. Land had become a possession to be owned.

Certainly, farmers are at the mercy of nature. Drought, floods, and insects and other pests destroy crops. More than any change in human history, farming most dramatically changed the way people lived with the land. With the development of farming, humans began to change the earth.

People are now able to damage the environment on a much larger scale. The environment in overpopulated parts of the world suffers as people try to feed their families. In nations such as Haiti and the Philippines, poor rural people have been driven into the hills and mountains to farm. The forests have been cut. This has created soil erosion that is ruining the land for farming. The eroded soil flows into streams and rivers. It is then carried into the ocean along the coast where it smothers young fish and other ocean life that people also depend on for food. Poor farming practices can harm more than just the land.

In the industrial world, modern agriculture supports an enormous human population. Modern farming allows a very small number of farmers to feed much of the world. Now we have machines that do much of the farmwork for us. This leaves the rest of us in the cities to do other things. We pay a price for this luxury. Modern agriculture is big business. It relies on chemicals and lots of energy. Chemical pesticides and fertilizers damage soils, pollute water, threaten wildlife, and may even be a danger to human health.

People in industrialized countries are referred to as consumers. We are consumers of the world's resources that are used to make all of the things that we buy. All of this consuming has put a strain on the world's natural resources and ecosystems. We buy and throw away a tremendous amount of material. We are encouraged to do this by our society, through advertising, movies, and television. We are encouraged to do this because it is the way that we are used to living. It is a part of our culture.

The earth cannot possibly sustain this trend. We cannot produce more and more things when we have a limited amount of resources. As the human population increases, there will be more people wanting more things. This cannot go on forever. We are already suffering from the consequences of this behavior. Cities have high population densities. This means that there are a lot of people living in a small area. As the population increases, more land is used and paved, leaving less land for wildlife.

People in cities need food, so farmers push their land harder to grow more food. To do this, they use more chemicals and more energy. These chemicals and the pollution from our cities do not stay in one place. DDT, a pesticide used in the 1950s and 1960s, has been found in the fat of penguins in Anarctica. Our impact on the environment has become global. We are now trying to deal with worldwide problems such as global warming, acid rain, ozone depletion, ocean pollution, and extinction.

You Solve It

Antarctica is the world's coldest, driest, highest, and windiest continent. Though it has the most remote ecosystems, it is being invaded by pollutants, tourists looking for adventure, and scientists looking to explore earth's last frontier.

1. Prepare a report summarizing information about Antarctica's geography and history.
2. Identify and describe the biotic and abiotic parts of this unusual ecosystem.
3. Find out who is scrambling to get their hands on the treasures of this continent and why. Find out what effects this human invasion is having on the ecosystems of this continent.
4. What do you think should be done to protect this ecosystem? Should we forbid people from setting foot on this continent, or should a more modest approach be sought? Outline your approach to saving this wilderness. Then look up information about the Antarctic Treaty that was signed in 1959 and revised again in 1991. How does your approach compare with this treaty? Do you think that this treaty is doing enough to protect this ecosystem? Explain.

SECTION REVIEW

1. Can you think of some additional human-controlled ecosystems other than the ones in the text?
2. Are human-controlled ecosystems entirely in human control, or does nature play a part in them, too? Explain.
3. Describe the impact of human tools on ecosystems.

FOR DISCUSSION

Why do we say that few parts of the earth remain untouched by the activities of humans?

Case Study

Tracking a Shadowy Urban Group

John Hadidian knew that his subject—adult male number 85—was nearby. The strong radio signal he was hearing through his headphones told him so. Following the radio signal, Hadidian walked quietly to peer into a dimly lit alley near downtown Washington, D.C. Hadidian, a special kind of detective, had followed this signal all over the city. It led him from a known hideout in a quiet neighborhood across town to this spot behind a fast-food restaurant. Now his subject was looking for leftovers in an open garbage bin, pushing aside everything that was not food.

John Hadidian had tracked this urban dweller for several months, recording his movements and learning his habits. With a portable radio receiver, Hadidian could find him from a distance at almost any time of the day or night. Hadidian also knew that adult male number 85 was part of a secretive group whose habits were unknown to most of the neighborhood people.

John Hadidian is not a high-tech police detective chasing a thief. He's a wildlife biologist with the National Park Service. Working for the U.S. government, he studies how urban raccoons live. Hadidian has followed hundreds of raccoons in the nation's capital. He puts collars with tiny radio transmitters on raccoons and tracks the animals by following the radio signals. He says that miniature radio transmitters have greatly helped biologists learn about urban animals. "We're studying raccoons," Hadidian says, "because we know they live in the city with us, but we know very little about what their lives are like."

For the past six years, Hadidian has tracked raccoons in the streets, alleys, and storm drains of Washington. Because the raccoons sleep during the day, Hadidian has spent many nights walking city streets wearing his headphones and carrying his tracking antenna. The job is especially difficult because raccoons are clever and avoid human contact. Raccoons have become very good at getting around the city unseen by people. Because of this, Hadidian says, our urban raccoon population is far greater than most of us realize. "We rarely see them, but raccoons are all around us," he says.

How many raccoons live in Washington? John Hadidian has estimated their number, based on his studies. He claims that there are between 15,000 and 20,000 raccoons in Washington, D.C. Hadidian says there can be up to 20 times more raccoons per acre in a city than in an acre of forest or farmland.

Raccoon Hangouts

Hadidian and his team of wildlife biologists can track individual raccoons on their daily rounds. This is because each radio transmitter on an animal wearing a collar sends its own separate signal. Using radio tracking, the team has found 1,900 raccoon dens or "hangouts" in Washington. The list of habitats shows the raccoon's ability to make a home in almost any spot. Favorite raccoon hideaways have been found in unused chimneys, in attics, in construction debris, and in sewers, garages, tree hollows, and shrub thickets.

One surprise about urban raccoons is how mobile they are. Hadidian's coworker, John Manski, tracked one male raccoon nearly 50 miles in one long trip. The animal went through the entire city and finally to nearby Reston, Virginia. This active raccoon swam across the Potomac River and used the city's network of alleys and underground storm drains.

Raccoons' success at city life can also be seen in their urban diet. Biologists know that rural raccoons hunt rodents and worms, but raccoons also eat berries and nuts. In farming areas, raccoons also eat corn and other crops. Their urban diet is even more varied. Raccoons eat garbage and snack on every kind of leftover fast food, from hamburgers to pizza crust. They are equally happy to steal seed from bird feeders or eat worms, fruit, frogs, insects, or mice.

Manski says raccoons in the city really stretch their varied diet. Analysis of their droppings—or scats—certainly shows that they aren't too picky. Hadidian and Manski have identified 34 separate species of plants in the raccoon diet. The plant food ranges from acorns to raspberries and crabgrass.

It's no surprise that a raccoon's diet varies dramatically with the seasons. Plants—especially fruits, berries, and nuts—are always a favorite. But in the fall, wild-plant foods make up a much larger share of the raccoon diet. These foods are more plentiful in autumn than at other times of the year.

Hadidian and Manski also learned that raccoons eat small rodents, such as mice, mainly in the spring. Some 47 percent of all raccoon scats contained the remains of small mammals in the months of April through June. But only 9 percent of the scats contained mammal remains between October and December, when nuts and berries are the most plentiful.

Raccoons in the city also scavenge more food in the winter. Evidence like wrappers, aluminum foil, or other telltale signs were found in 50 percent of all the raccoon scats collected in the months of January through March. This shows that food was scavenged from human trash. But only 19 percent of the scats collected between October and December had human refuse.

Hadidian's radio tracking also helped him learn that individual raccoons sometimes go a long way to get a favorite treat. Hadidian recalls one raccoon that he tracked for two consecutive years. This individual changed his normal behavior pattern at almost precisely the same time each year. He moved from his comfortable home in an unused chimney in one neighborhood to a tree den across town. The raccoon camped out for a week in this tree, which was in the National Zoo. "The second year," Hadidian says, "I realized that this raccoon had returned to await the ripening of persimmons at a particular tree he favored. He might well have been making the trip every year for a long time."

Teens Track Raccoons

John Hadidian and Dave Smith, a Park Ranger at the Rock Creek Park in Washington, conduct a summer program for some two dozen urban youths. During this "Junior Ranger" program, Hadidian and Smith let the young people use the radio-tracking equipment in the park and find raccoon dens for themselves.

Using the tracking equipment, the Junior Rangers can almost always find a raccoon nearby. The animal is often sleeping in a hollow of a tree or sometimes just lying lazily in full view across a large branch.

Smith says the experience helps the youth get a firsthand feel for the excitement of tracking animals in their natural habitats. The tracking work also creates a lot of interest in the interactions of different species within an ecosystem. "Plus," Smith says, "the tracking equipment is fun to use. When you put it on it makes you look and feel like some sort of secret agent."

2 ECOSYSTEM INTERACTIONS

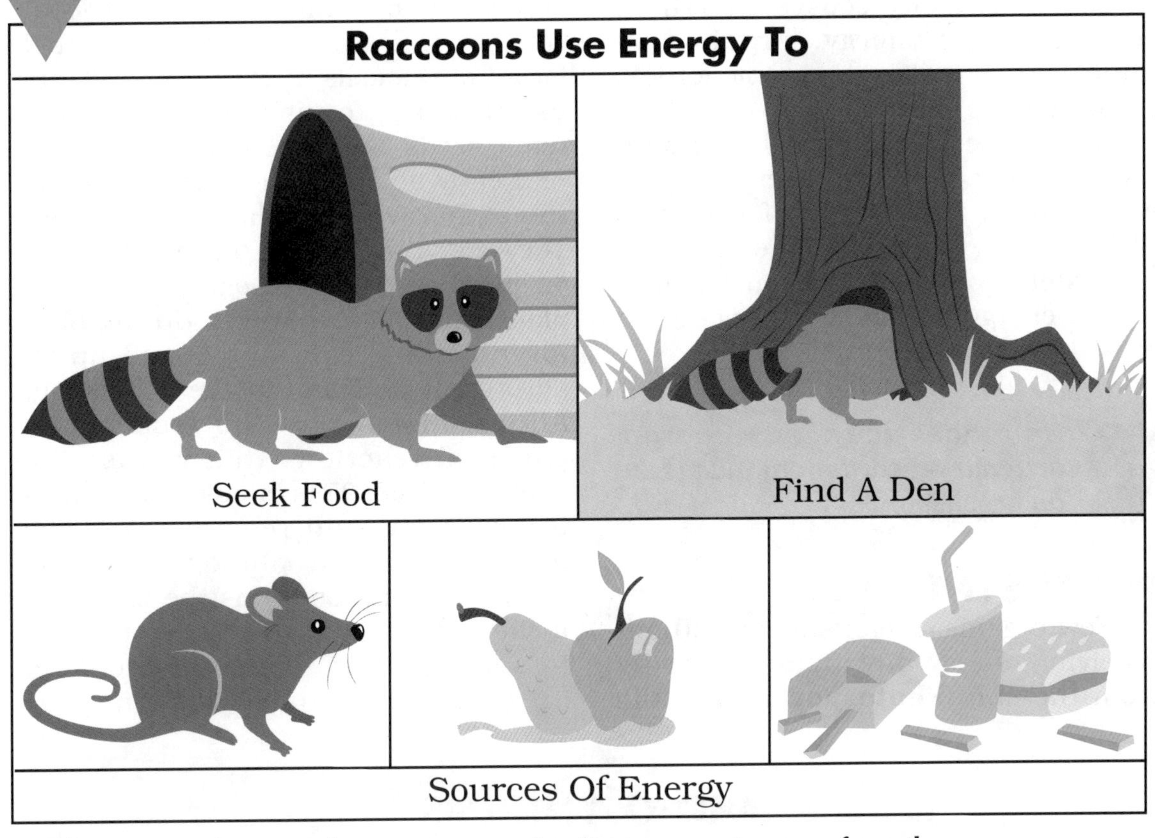

All living organisms need an energy supply. Raccoons get energy from the environment by consuming a wide variety of foods.

2.1 Energy Flow in Ecosystems

John Hadidian and his coworkers have found that raccoons can live in much more crowded populations in cities than in the wilderness. In Washington D.C., an area of about 100 acres can contain 42 raccoons. (A 100-acre area can hold 75 football fields.) There might be only 4 to 8 raccoons in a wilderness area roughly equal in size. People usually think that wilderness is good for wildlife. Why, then, are there so many raccoons in a city ecosystem?

Part of the answer, as you have read in the case study, is that raccoons have learned to live close to humans. Not all wild creatures can manage to do this. Another part of the answer is that the city ecosystem and a wilderness ecosystem have very different energy supplies. This is because humans transport much energy—in the form of fuel, food, and electricity—to a city. This energy may be transported from great distances to the city. Some of the human food is wasted and can be used by animals. Some of the fuel and electricity changes to heat, which warms parts of the city used by raccoons. The city ecosystem, then, contains some energy, supplied by humans, that makes it possible for lots of raccoons to survive.

A wilderness ecosystem has a supply of energy, too. This energy supply comes from the energy of the sun. With few exceptions, the natural ecosystems on earth rely on the sun for the energy that all living organisms must have to survive.

How does the sun's energy actually reach the organisms in an ecosystem?

PHOTOSYNTHESIS

Few organisms can survive without receiving energy directly or indirectly from the sun. **Photosynthesis** is the process by which green plants and algae absorb solar energy. They use this energy to combine carbon dioxide and water to form glucose, their food. Glucose is a simple sugar and a fuel for all of the cells in organisms. Plants use this fuel to make leaves, stems, roots, seeds, and other parts.

Nearly all other organisms must get their energy from plants (or algae). They do so by feeding on the plants or on plant-eaters. The solar energy that plants and algae harness during photosynthesis is constantly used up by all organisms living in the ecosystem. All organisms carry on respiration, chemical reactions that release the energy in food. This enables organisms to live, grow, and reproduce. As the food produced by photosynthesis is used up, more food must be produced if the organisms in the ecosystem are to survive. Thus, the solar energy in an ecosystem must constantly be

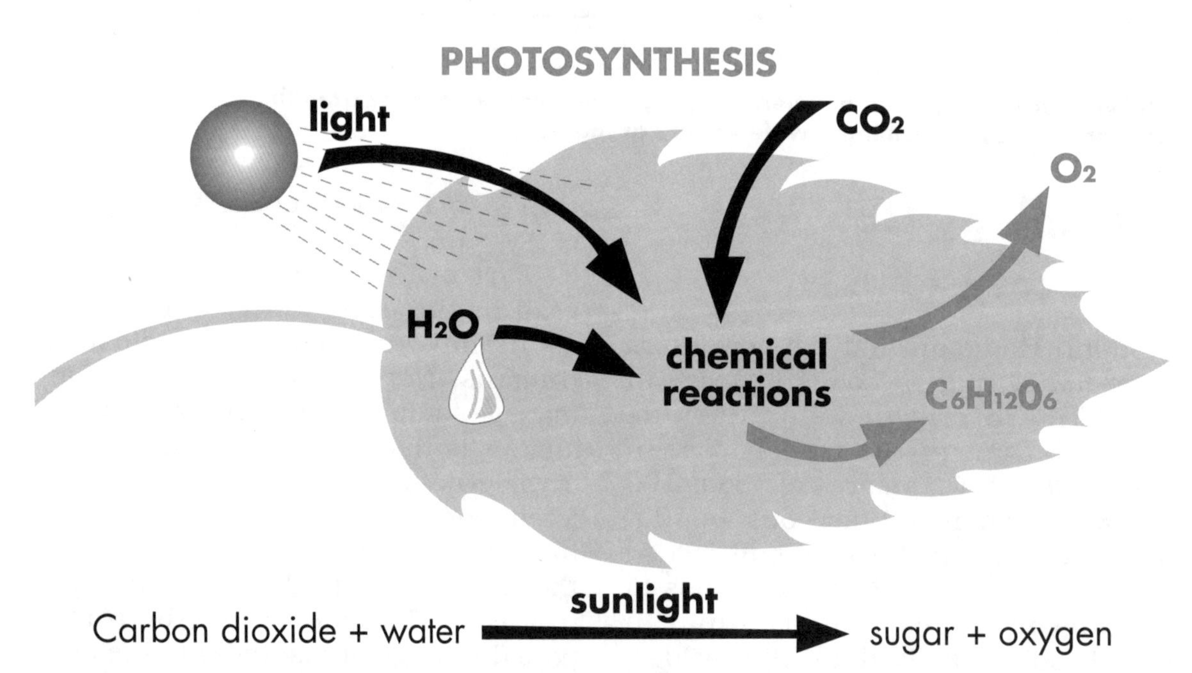

In most ecosystems, the process of photosynthesis is the source of the energy supplies for the living organisms, including people.

replaced. Ecologists say that energy from the sun flows through an ecosystem.

ENERGY FLOW

A tree in a city park uses energy from the sun to make its own food through photosynthesis. The caterpillar eats the tree's leaves, and the bird eats the caterpillar. A raccoon may eat the fruit of the tree. A mushroom gets food from a dead or decaying tree.

In ecosystems, organisms can be grouped according to the way in which they obtain energy. The tree, mushroom, caterpillar, raccoon, and bird each represent different ways of getting food in an ecosystem.

Producers Producers make their own food by photosynthesis. The most common producers on land are green plants such as trees and grasses. The most common producers in water are called **phytoplankton**, which are microscopic organisms. Examples of these organisms include Volvox, a type of algae, and Euglena, a one-celled organism.

A few producers, such as bacteria in deep ocean vents, make their own food by **chemosynthesis**. In chemosynthesis, the bacteria use chemical reactions to convert simple compounds, such as ammonia, to other simple compounds and to release energy at the same time. This energy is used to make sugar from carbon dioxide and water.

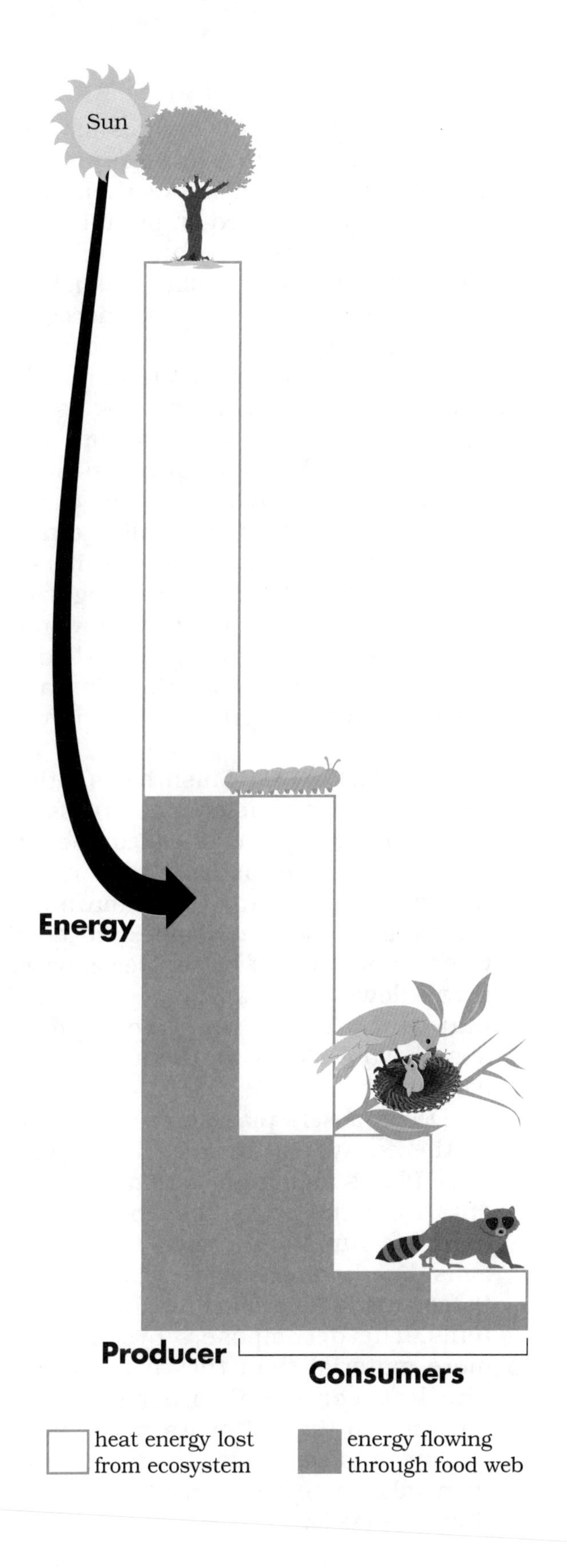

Energy flows from the sun to plants and then to animals. At each step, some energy is lost, some is used to maintain life functions, and some is used for growth and reproduction.

Consumers All organisms that are not producers are **consumers.** Consumers cannot make their own food so they must feed on producers or other consumers to obtain energy. Most consumers are animals, such as worms, insects, pigeons, and raccoons.

Consumers are grouped according to what they eat. **Herbivores** are consumers that eat only producers, usually green plants. **Carnivores** are consumers that eat only animals including herbivores and other carnivores. Other consumers, called **omnivores**, eat both plants and animals. Into which of these two categories would you place the raccoons that were described in the case study? What evidence did John Hadidian and his coworkers present to support your answer?

Decomposers Mushrooms, other fungi, and many bacteria are an essential group of consumers called **decomposers**. Decomposers break down waste material and the remains of dead organisms into simple nutrients. Chemicals given off by the decomposer break down the dead plant or animal material. Then the decomposer absorbs the simple nutrients through its body.

Decomposers play an essential role in the ecosystem by recycling nutrients. Plants and animals remove nutrients from the ecosystem to carry on their life functions. When the organisms die, decomposers recycle many of those nutrients back into the ecosystems. The decomposers break down more material than they can absorb. The leftover simple nutrients are returned to the soil or water in which the decomposers live. Producers can then take in these nutrients and continue the cycle.

FOOD CHAINS AND FOOD WEBS

It is known that mice in the wild feed on grains, berries, and seeds. As you have read in the case study, raccoons feed on mice. In an ecosystem, the sequence of organisms that eat one another is called a **food chain**. The diagram on p. 25 shows energy flow in a food chain from Rock Creek Park in Washington, D.C.

The park food chain seems relatively simple. However, you may have realized that the feeding relationships within the park ecosystem are not fully described by this single food chain. Consumers feed on or are fed on by more than one organism in the ecosystem. For example, the raccoon feeds not only on mice but on berries and worms as well. The red fox and raccoon both feed on the mice.

In many ecosystems, food chains are only a part of a more complex system of energy relationships. Usually, energy flows through a number of interconnected food chains. A group of interconnected food chains is called a **food web**. What are some of the food chains that make up the food web shown on p. 27?

Each species in a food web can be thought of as one link. The more complex the web, the less likely that the elimination of one link will effect the survival of the other organisms in that web.

You may have heard the expression "A chain is only as strong as its weakest link." The same can be said of a food chain. A food chain often contains a certain species whose elimination would mean the destruction of the entire chain. For example, the importance to the food chain of the **zooplankton**, which are tiny floating

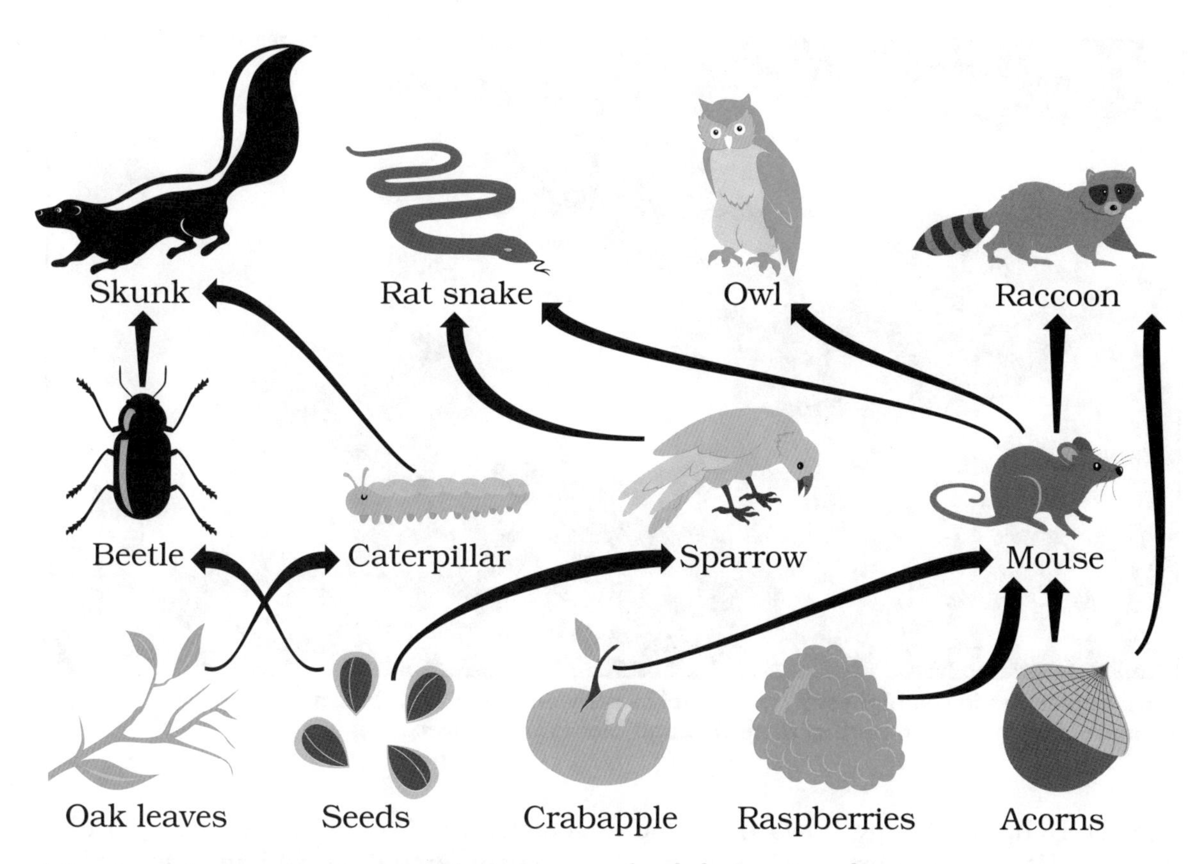

A food web in an ecosystem may include many food chains. Use the arrows to trace the food chains in this food web.

animals, in Flathead Lake, Montana, was demonstrated when their populations were almost destroyed.

In the 1960s and 1970s, wildlife agents released freshwater shrimp, a nonnative animal, into the streams that lead into Flathead Lake. The shrimp were meant to increase the food supply of sport fish in the streams and thus increase the supply of fish for the sport fishers. While this goal was successful, there was also an unfortunate effect.

The shrimp made their way into the lake where they took over the habitat of the native zooplankton, nearly wiping them out. Because the native zooplankton constituted the only food source of the lake's kokanee salmon, their population also declined sharply. As a result, the salmon population also declined sharply. Soon after, the local population of bald eagles, already an endangered species, was destroyed because the eagles were deprived of their main food source, the kokanee salmon.

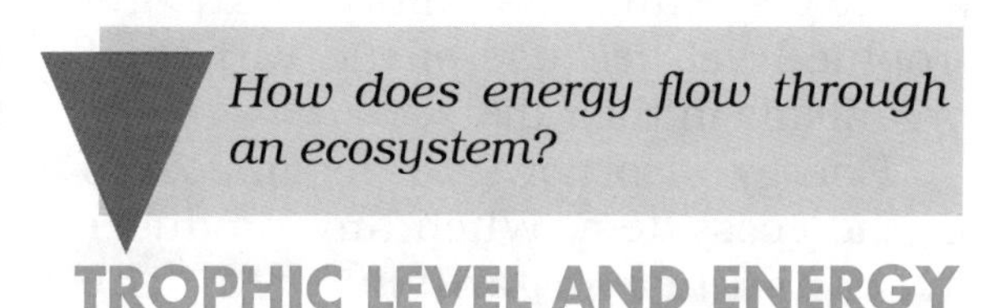

How does energy flow through an ecosystem?

TROPHIC LEVEL AND ENERGY

The location of an organism along a food chain is called its **trophic level**. Recall that a tree, which manufactures its own food, is a producer. A producer is always in the first trophic level in

Bald eagles are in the highest trophic level in their environment. Harmful chemicals from pollution affect them through biomagnification.

the food chain. Energy flows from the sun to the producer level in most ecosystems.

The second level contains the consumers that eat the producers—the herbivores, such as a caterpillar. In the third trophic level are carnivores, such as birds that feed only on caterpillars. The herbivore-eating carnivores are, in turn, eaten by other carnivorous consumers, such as hawks. Carnivores and omnivores, such as raccoons and red foxes, can be at more than one trophic level because of the variety of organisms in their diets.

Energy also flows to decomposers in the ecosystem. When any producer or consumer organism dies, the decomposer organisms break it down and obtain energy. Why can decomposers be at any trophic level except the first one?

Species in the higher trophic levels often act as an "alarm system" for their ecosystem. A decline in their populations can signal a problem at lower levels. One such problem was discovered in the ocean ecosystem when large amounts of industrial wastes, especially mercury, were dumped into the ocean.

When tiny phytoplankton absorbed the mercury from the water, the concentration in them was not very high. Then the zooplankton ate the mercury-tainted phytoplankton. Because they ate a large number of the phytoplankton, the concentration of mercury in the zooplankton was higher than in the phytoplankton. Zooplankton is the primary food source of anchovy, which is a food of tuna.

When the tuna ate large numbers of anchovies, the level of mercury in their system became so high that people would be harmed if they ate the tuna. So you can see how a small amount of mercury pollution at the first trophic level becomes more concentrated, or magnified, as it makes its way up the food chain. This is called **biomagnification**.

Few food chains have more than five trophic levels. The length of the food chain is limited because, at each level, some energy is lost. Energy is used up by each organism in order to carry on its life functions, including digestion, circulation, respiration, growth, and reproduction.

Energy also is lost as heat from the bodies of the organism. You lose heat from your body when you perspire. More energy is lost as energy flows from level to level. This means that at higher trophic levels, less energy is available for animals. In fact, only 10 percent of the energy available at each level is transferred to the next level. The amount of energy transferred from level to level is illustrated by the diagram shown on p. 25.

You Solve It

DDT is a pesticide that has caused widespread damage to the environment since it was first used in World War II. It is so effective in killing insects such as mosquitos that it was applied to fields and bodies of water in a very dilute concentration, about 0.00005 parts per million, or ppm.

1. What does *parts per million* mean? What do *ppt* and *ppb* mean? Why do you think these units of concentration are used for solutions of pesticides like DDT?
2. A typical food chain along the shore of Long Island may consist of plankton, clams, and gulls. When these organisms were analyzed for DDT content, the following results were obtained: plankton, 0.04 ppm; clams, 0.42 ppm; and gulls, 18.5 ppm. How can you explain these concentrations when the water itself contained only 0.00005 ppm DDT?
3. Because it is so detrimental to the environment, it is not surprising that DDT was banned in the United States in the early 1970s. What may be surprising to you, however, is that DDT is still being produced in the U.S. for export. Find out how much DDT is still being produced in the United States. for export. What countries are importing DDT and what are they using it for? Identify the ecological problems that are being caused by the use of DDT in these countries. Do you think the benefits of its use in these countries outweigh the damage?
4. There is a very popular environmental phrase "Think Globally—Act Locally!" Should we be concerned about the damage being done by DDT in other countries? What is our responsibility as a nation to stop the use and global damage caused by DDT? What can be done to stop the global spread of DDT?

How do the biotic factors in an ecosystem in your area interact?

The biotic factors of an ecosystem interact in complex ways. The biotic factors can be classified as producers, consumers, and decomposers. The consumers can be further classified as herbivores, carnivores, omnivores, or scavengers. In this activity you will study and classify the organisms in a ecosystem near your school or home. Using this information you will then define how these organisms interact and summarize this interaction by diagraming a food chain.

Materials

hand lens
meter stick
journal
string
markers
stakes
hand spade or shovel
white paper or cloth

Procedure

Part 1

1. Select an ecosystem near your school or home. Your teacher will recommend the approximate size for the ecosystem. Use the meter stick to measure off the appropriate amount of land.
2. Use the string and stakes to mark the land you select.
3. Draw a map of the area you staked off. You may want to include the abiotic factors such as large rocks or streams to complete your map.

Part 2

1. Identify the biotic factors present in your ecosystem. Use the chart provided by your teacher to tally and describe the organisms you find. Then note which ones are producers, consumers, or decomposers. Sketch each organism in your journal or notebook.

ORGANISMS FOUND IDENTITY TALLY	PRODUCER	CONSUMER H* C* O* S*	DECOMPOSER
1			
2			
3			
4			

* H=herbivore O=omnivore
C=carnivore S=scavenger

2. Scan your area and identify the prominent organisms, such as trees, bushes, flowers, and grasses. Record the names of the organisms you identify. Look more carefully for the less prominent organisms.
3. Examine a portion of the soil. Place a handful of soil on a piece of white paper. Sort through this to locate any organisms that may be present.

4. Place a piece of white cloth or paper under bushes and shrubs. Gently shake them to see if any organisms fall onto the cloth or paper.
5. Look under leaf litter, logs, and rocks to see if there are any organisms living there.
6. Look for evidence that other organisms may occupy or have occupied this area in the recent past. Record this evidence in the chart provided by your teacher.

	EVIDENCE OR SIGNS OF OTHER ORGANISMS	POSSIBLE IDENTITY OF ORGANISM
1		
2	(i.e. scat, feather, burrows, decomposed matter)	
3		
4		

7. Spend one class period observing your ecosystem from about 10 feet and 100 feet away. Determine if other organisms such as insects, birds, or squirrels enter your area. Include these organisms in your chart.

Part 3

1. Study the information in your chart. Think about interactions among the organisms you discovered in your ecosystem.
2. From the evidence you gathered, draw food chains to represent the interactions among the organisms. Use arrows to show which factors affect each other.

Conclusions

1. What relationships exist between the organisms you observed? What evidence did you use to establish these relationships?
2. Are there more producers or consumers in your ecosystem? Why do you think this is so?
3. Did you see any decomposers in your ecosystem? If not, did you see any evidence of the presence of decomposers? If so, what?
4. Did you find any evidence that other animals may have visited your ecosystem at other times during the day or night?

For Discussion

1. Predict what would happen if you removed one of the producers from your ecosystem.
2. Predict what would happen if you added a specific consumer to your ecosystem.
3. Study and compare the charts of your classmates. How can you account for any differences that were recorded.

Extensions

1. Study the maps and food chains of your classmates. Are there any adjoining or overlapping ecosystems? Can you combine the food chains into a food web?
2. Obtain reference books from your school or local library. Use these books to help you identify the organisms that you found in your ecosystem.

SECTION REVIEW

1. Make a list of organisms in or around the school building that perform photosynthesis. Are any of these organisms part of a food chain?
2. Make a list of everything you ate yesterday. Identify the organisms that the food came from. Then put each food in one of these categories: producers, herbivores, carnivores, or decomposers.
3. List the components and draw a picture of a possible four-level food chain of organisms in your neighborhood.
4. Based on the information provided in the case study, draw a food web to show the feeding relationships that involve the raccoons.
5. Draw an energy diagram, similar to the one on p. 25, on a piece of paper. Put the organisms shown in the food web on p. 27 into your diagram.

FOR DISCUSSION

1. How does the supply of energy in an ecosystem change as you "move up the food chain?"
2. Why is photosynthesis so important to humans' survival?
3. If there were no decomposers in an ecosystem, what would happen to all the plant and animal remains?

2.2 Interactions Among Populations

You may recall that different species can share the same habitat. Because these organisms share a common environment, they often interact. The outcome of the interactions of two species can harm or help one or both of them.

Do ecosystems need predators?

PREDATION

One interaction in which one species benefits from another is called **predation**. In predation, an organism of one species, called the *predator*, feeds on all or part of an organism of another species. The species being eaten is called the *prey*.

A dramatic and obvious example of predation is a lioness killing and eating an antelope. However, a raccoon eating a worm or insect is also an example of predation. Animals can be both predators and prey. A crayfish in Rock Creek in Washington, D.C., is a predator when it eats a stream insect. The crayfish might later become the prey of a raccoon hunting in Rock Creek.

Predation and Energy Flow The predator-prey relationships in an ecosystem are an essential part of the

This lion, a top predator in its ecosystem, has captured an antelope for food. Lions use up much energy when they hunt.

flow of energy. Some predator-prey interactions consume much energy, while others consume far less. The act of hunting and capturing prey can use up large stores of a predator's energy. Some predators, such as tiny shrews, need huge amounts of energy to carry on their life functions. Therefore, shrews spend almost all their time hunting and eating to take in enough energy.

Other predators, such as snakes, expend little energy in their daily life functions. Unlike shrews, snakes use little of their internal energy to maintain a steady body temperature. Instead, they use energy from the sun by lying on sun-warmed rocks and bathe themselves in sunlight to stay warm. This reduced need for energy results in a less frequent need to hunt. A snake may hunt only once a month, devouring only one small mammal, such as a mouse. The snake will then slowly digest the mouse over several days. The slow digestion process also requires less energy and, thus, less food.

In some ecosystems, predator-prey interactions are a key to maintaining stable populations. Suppose there was an ecosystem in which the lynx, a kind of wild cat, was the only predator of the rabbit, which fed on the vegetation. If the lynxes were eliminated, the rabbits would reproduce indefinitely. Eventually, there would be more rabbits than the vegetation in the ecosystem could support. The rabbits could spread out to other ecosystems, which they might destroy. If there were no other ecosystem to move on to, then

This snake is a predator that consumes rodents. Interactions like this one between a predator and prey are a key part of most ecosystems.

the rabbit population would eventually die of starvation.

Thus, when the predator species is somehow limited or eliminated from an ecosystem, the prey species may increase tremendously. This increase of one species could result in another species being crowded out of the ecosystem or the producer of the ecosystem being destroyed through overgrazing.

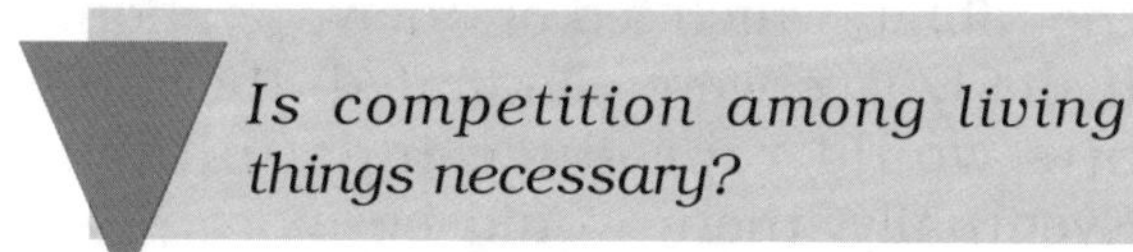

Is competition among living things necessary?

COMPETITION

When the basic needs or activities of different species overlap, the different species are forced to interact. **Competition** is the interaction of two species when they both need to obtain the same limited supply of a resource.

During competition, one species may limit or prevent the other species from using a certain resource within the ecosystem. The freshwater shrimp that were introduced into Flathead Lake may have been eating the same foods as the native zooplankton (see *Food Chains and Food Webs* pp. 26-27.) If the food supply in the lake was not enough to support both the shrimp and the zooplankton, then one of these two groups of consumers would vanish. This is an example of how competition between different kinds of animals can cause a population to disappear.

Sometimes, two species use a resource at different times so that it can be shared. This is one way to reduce or avoid competition. Also, the

The starling (left) and the bluebird (right) compete for nesting space, a limited resource.

more common a shared resource is, the more easily the two species can share it. If the mice in Rock Creek Park are very abundant, for example, then the raccoons and red foxes need not compete for this food source. However, when less of a resource is available, the two species must compete for its use. This might happen if, for example, both raccoons and foxes sought ripe persimmons to eat.

Competition can also take place between the members of the same species. John Hadidian and his coworkers in Rock Creek Park, Washington, D.C., reported that raccoons did not seem to compete for food or resting places. His explanation is that both food and resting places are abundant resources for city raccoons. In a wilderness, however, these resources may be limited in quantity. Perhaps competition among wilderness raccoons is one of the reasons why raccoons in a wilderness are so much more spread out than they are in a city.

SECTION REVIEW

1. Do you think that every food web must contain predators? Why or why not? Are there any examples of predators in your neighborhood or near your school? If so, what are they?
2. Are humans predators? Give examples of humans acting as predators in an ecosystem. Then compare your examples with those of a partner.
3. If the people of Washington, D.C., could prevent raccoons from eating most of the food that gets discarded, what effect might this have on interactions among raccoons in the neighborhoods?
4. Can you think of examples of organisms that compete with humans for food?

Trees growing on these peaks in the Great Smoky Mountains are limited by the local conditions such as elevation, temperature, and depth of soil.

2.3 Limiting Factors

You may wonder why every species doesn't spread into all kinds of different ecosystems. This is because each species has a limited range of conditions that it can tolerate. A species may have a wide range of tolerance for some conditions and a small range of tolerance for others. The more factors for which a species has a wide range of tolerance, the more varied the environments it can live in.

The components of an ecosystem determine whether a certain species can survive there. Some components are called limiting factors because population size of a species can be limited if too much or too little of that component exists in the ecosystem. This is true even if all the other components of the ecosystem are within the species' range of tolerance. These components include length of daylight, temperature range, humidity, precipitation, winds, water currents, water salinity (the amount of salt dissolved in water), amount of space, and soil or water nutrients.

Once a condition reaches a point past the tolerance range of an organism, the organism will die, often abruptly. This can happen when an ecosystem changes naturally. Pollution also changes conditions in ecosystems, causing the death of organisms.

For example, many spruce trees in eastern North America began to die very quickly at about the same time in different areas. These areas included parts of eastern Canada, New England, and New York's Adirondack mountains. Air pollutants in these areas had been building up over time. The pollutants played a part in making acid rain. The acid rain reduced the spruce trees' ability to survive. Scientists studying the problem learned that acid rain limited the trees' tolerance for cold. Acid rain, therefore, became a limiting factor in the survival of spruce trees. The biotic parts of ecosystems can also be limiting factors.

Some animals, such as the koala are limited by the kinds of foods they can eat. Koalas can eat only the leaves and bark of the eucalyptus tree. Furthermore, of the 365 species of eucalyptus that grow in Australia, the koala eats from only 20 of those species.

The koala's diet is one of its limiting factors.

Other animals, such as the raccoon, have few food limits. They can survive under a wide range of environmental conditions, from wilderness to an urban park, partly because they are able to find and eat a tremendous variety of foods. The case study showed that they eat anything from small insects to discarded fried chicken.

SECTION REVIEW

1. What are some limiting factors of humans?
2. Suppose that chimneys, new storm sewers, and other parts of the urban landscape were built to exclude raccoons from using them. What might happen to a city raccoon population?
3. A limiting factor for a shrew population is the availability of their food source—worms. Suppose in one season there is a massive increase in the worm population. Hypothesize about its effect on the shrew population.

Lab Study

How does water temperature affect the rate of photosynthesis of an *Elodea* plant?

Limiting factors in an ecosystem restrict the ability of an organism to grow and survive. When any limiting factor is below a critical minimum level or above a critical maximum level, an organism cannot survive. The tolerance of different factors varies with species. In this activity, you will determine the minimum and maximum water temperature for an *Elodea* plant to survive based upon its rate of photosynthesis at various water temperatures.

Materials

4 test tubes
4 1-hole stoppers
4 flexible straws
4 beakers of water
***Elodea* plants**
hot plate with beaker of water
ice water bath
0.5% sodium bicarbonate solution
4 thermometers

Procedure

1. Work in a group with three other students. Each student in the group is to perform Steps 2 through 6 at the same time.
2. Insert the curved end of the flexible straw into the hole in the stopper. The end of the straw should be flush with the bottom of the stopper.
3. Fill the test tube with the sodium bicarbonate solution. Place the thermometer in the test tube. Use the hot-water bath or the ice bath, whichever is appropriate, to adjust the temperature of the solution as directed below.

STUDENT	TEMPERATURE OF TEST TUBE (°C)
1	10°
2	20°
3	30°
4	40°

4. Study the diagram on page 39. Remove the thermometer and put a sprig of *Elodea* into the test tube. Place the stopper in the test tube.
5. Each student should move the apparatus to the same location near a sunny window. This will ensure as close to identical light conditions as possible for each test tube. Hold the free end of the straw below the surface of the water in a beaker as shown in the diagram.
6. Using a stopwatch or a clock with a second hand, count the number of bubbles released by the *Elodea* plant in one minute. Record this number in the chart provided by your teacher.

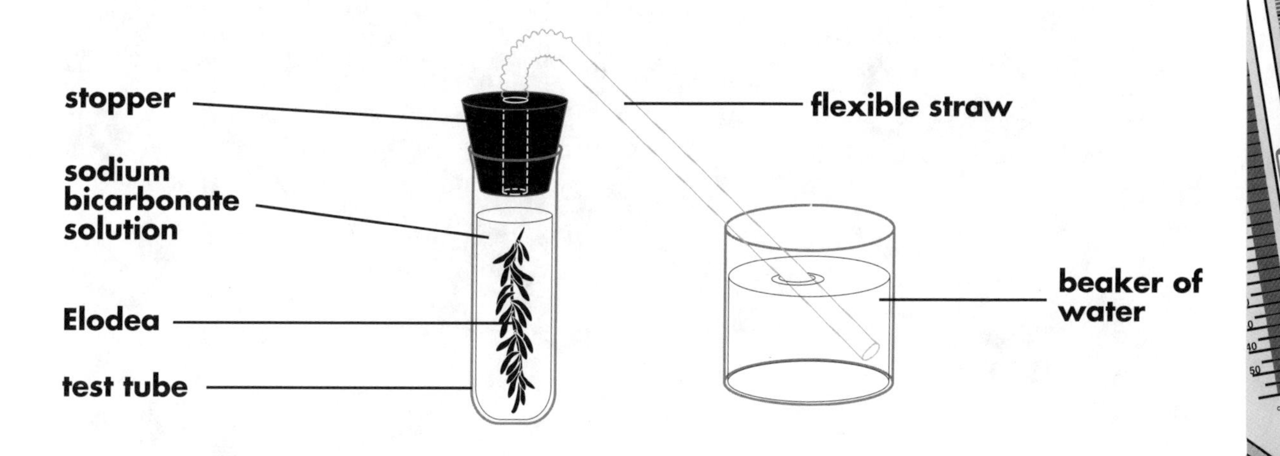

7. Repeat this measurement two more times and calculate the average number of bubbles produced in one minute.

Conclusions

1. Prepare a bar graph to show the average number of bubbles produced per minute at each of the four temperatures.
2. What is the relationship of water temperature to the rate of photosynthesis of the *Elodea* plant?
3. Could water temperature act as a limiting factor for an *Elodea* population? Explain your answer.

For Discussion

1. Why is it important that green plants in an aquatic environment continue to carry out photosynthesis?
2. What types of environmental conditions, both natural and human-made, might alter the water temperature of an aquatic ecosystem?

Extensions

1. Design an experiment to find out if the amount of dissolved carbon dioxide in an aquatic environment could be a limiting factor for an *Elodea* population.
2. Design an experiment to find out if the intensity of artificial light in an aquatic environment such as an aquarium could be a limiting factor for an *Elodea* population.

3 CYCLES IN ECOSYSTEMS

Ecosystems recycle matter, including water, minerals, and the chemical compounds that make up the bodies of living organisms. Without this recycling, you and other living things could not exist.

3.1 Recycling the Materials of Life

Do you recycle at home? Perhaps, like people in many communities, you recycle paper, glass, aluminum, and plastic. This helps businesses use less raw material when they make products.

Ecosystems recycle material, too. Ecosystems recycle the elements and chemical compounds that make up living organisms, soil, and rock. The elements that make up living and nonliving matter cannot be destroyed. The supply of these elements in ecosystems is finite, or limited.

Ecosystems do not recycle energy, however. In the last chapter you learned that energy from the sun flows through ecosystems. It is used up, for example, by the activities of organisms. Ecosystems need a steady supply of new energy. But the materials in an ecosystem must be reused over and over. Let's see how this works.

Like a circle, a material cycle has no beginning or end. Unlike a circle, a material cycle does not follow one simple path. A material may follow any number of pathways as it cycles through an ecosystem.

The pathways in a material cycle are actually events or processes. During an event or process, a material may be

- moved from one place to another (e.g., water flowing from a mountain top to an ocean)
- changed in physical condition (e.g., ice melting to form liquid water)
- combined with other materials (e.g., oxygen gas entering your lungs and then combining with other chemicals in your cells)

There are many material cycles that function in ecosystems. The best way to understand them is to look at examples. In this chapter you'll explore the cycles of water, carbon, nitrogen, oxygen, and a few minerals. Why do we try to understand these cycles? The answer is that they are essential to our lives.

Water Water covers most of the earth's surface. It is part of every ecosystem, even those that have dry weather. All living organisms contain roughly 80 to 90 percent water. The chemical reactions that maintain life need water. It is the main source of the element hydrogen in the compounds that build the body's cells.

Carbon You can't live without carbon. Your body's proteins, sugars, starches, fats, enzymes, and other chemicals all have carbon. This element makes up nearly one fifth of your body weight. All organisms need it, and ecosystems can not work unless carbon gets recycled.

Nitrogen, Oxygen Nitrogen is essential for life because it is a building block of proteins as well as DNA, the chemical that makes up your genes. Oxygen, like carbon, is part of the chemicals of life. It is also one of the most abundant elements on earth. In this chapter, you'll trace the movement of oxygen in an ecosystem through the pathways of the water and carbon cycles.

WATER CYCLE

Let's begin with the water cycle, for its pathways are familiar and easy to visualize. You are also familiar with the three physical forms that water may take in the environment. As a liquid, water occurs on the earth's surface in oceans, lakes, and rivers. Liquid water is in the soil and is trapped underground among rock layers. It is also in the air as clouds and fog.

As a solid, water occurs as ice. Most ice in the environment is at the earth's polar regions in the **Arctic** and the **Antarctic**. High mountain **glaciers** contain some ice as well. As a gas, water vapor occurs in the atmosphere.

Let's begin tracing the water cycle in the oceans, which hold most of the earth's water. Water may leave the ocean when it evaporates and enters the atmosphere. **Evaporation** is the process that changes a liquid into a gas. Sunlight provides the energy that is needed for this process to take place.

Water vapor may be moved around the atmosphere by winds. It may also change back into liquid form, which you can see as clouds, fog, or mist. This change is called **condensation** and happens when the vapor cools. If water in the atmosphere cools still further, it may form solid particles of ice or crystals of snow.

Eventually, water returns to the earth from the atmosphere as **precipitation**. Some of this water may fall directly into the ocean. This is an example of a short pathway in the cycle: ocean <——> atmosphere

Much precipitation, however, falls on land. At this point, water may follow several different pathways. Rain may soak into the soil and enter the **groundwater**. It may run off the land surface and enter lakes or rivers. Water in the soil may be taken up by plant roots and then remain within the bodies of living plants, animals, and microbes. Snow falling on high mountains may be packed into a glacier, where it may stay for thousands of years.

Water absorbed by plants can return directly to the atmosphere. Through the process called **transpiration**, plant cells in leaves release

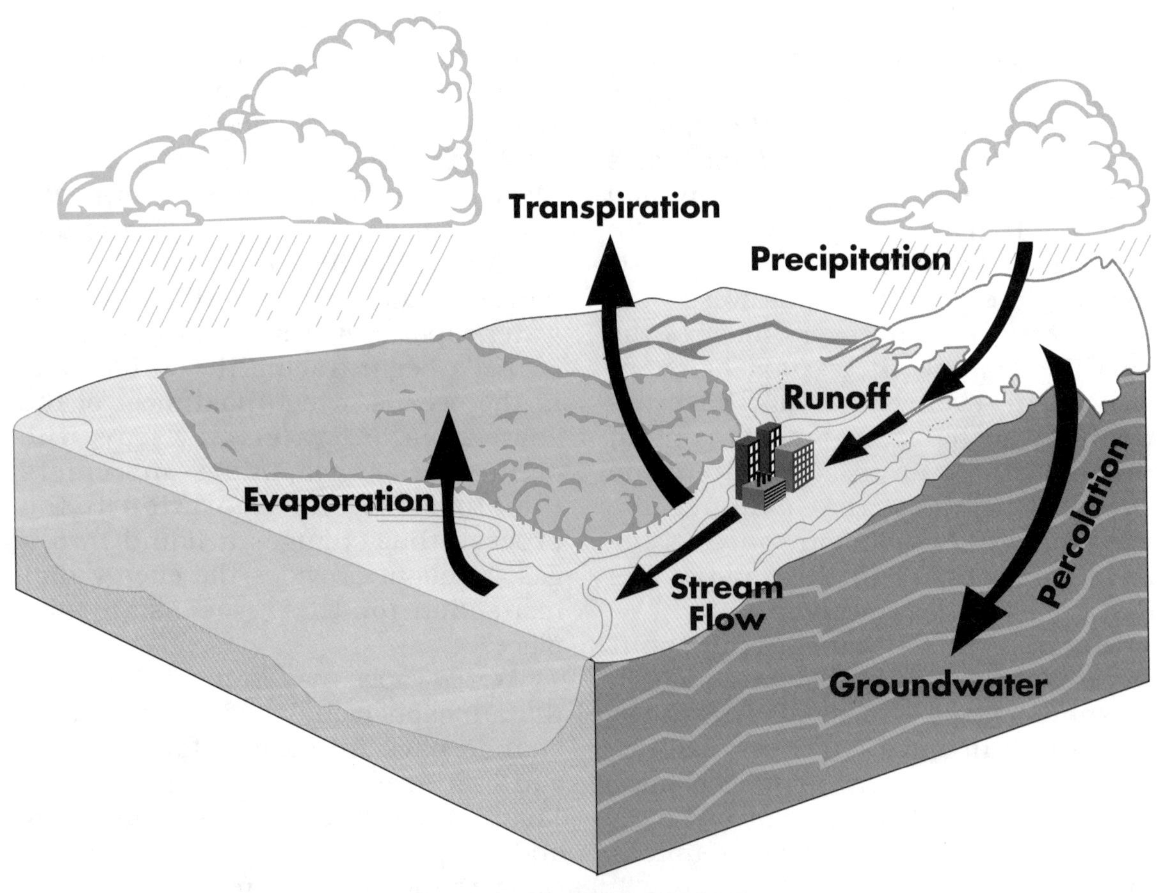

THE WATER CYCLE

The arrows indicate water cycle pathways between the atmosphere, land, and bodies of water. Percolation is the pathway water takes as it filters down through the soil. Ground water can flow and feed springs, rivers, and lakes.

Note the fungi growing on the side of this log. They look like tiny shelves stuck to the log. Fungi help break down the remains of organisms. Such decay releases water and carbon compounds into the ecosystem.

water into the air. This is an example of a longer pathway in the cycle: atmosphere ——> soil ——> plants —> —> atmosphere.

Water that enters plants may become part of the process of photosynthesis. You learned about photosynthesis in the previous chapter. At this point in the cycle, the plants use water as a raw material to make oxygen and sugar. Water is changed by the chemical reactions of photosynthesis. How, then, does the cycle continue? When a living plant or animal dies, the process of decay releases water back into the environment.

Sooner or later, all water returns to the ocean. How long this takes depends upon whether water enters living organisms, groundwater, or polar ice caps. The different pathways may take vastly different amounts of time.

Examine the diagram of the water cycle on p. 42. You can see that the events in the cycle include evaporation, precipitation, and transpiration. These events direct water along different pathways in the cycle. The ocean, the polar ice caps, and groundwater are places where water can remain for long periods of time. We refer to these parts of the cycle as reservoirs.

THE CARBON CYCLE

The carbon cycle is really several cycles that are connected to one another. Carbon cycles follow several different pathways in the environment. The time needed for carbon to com-

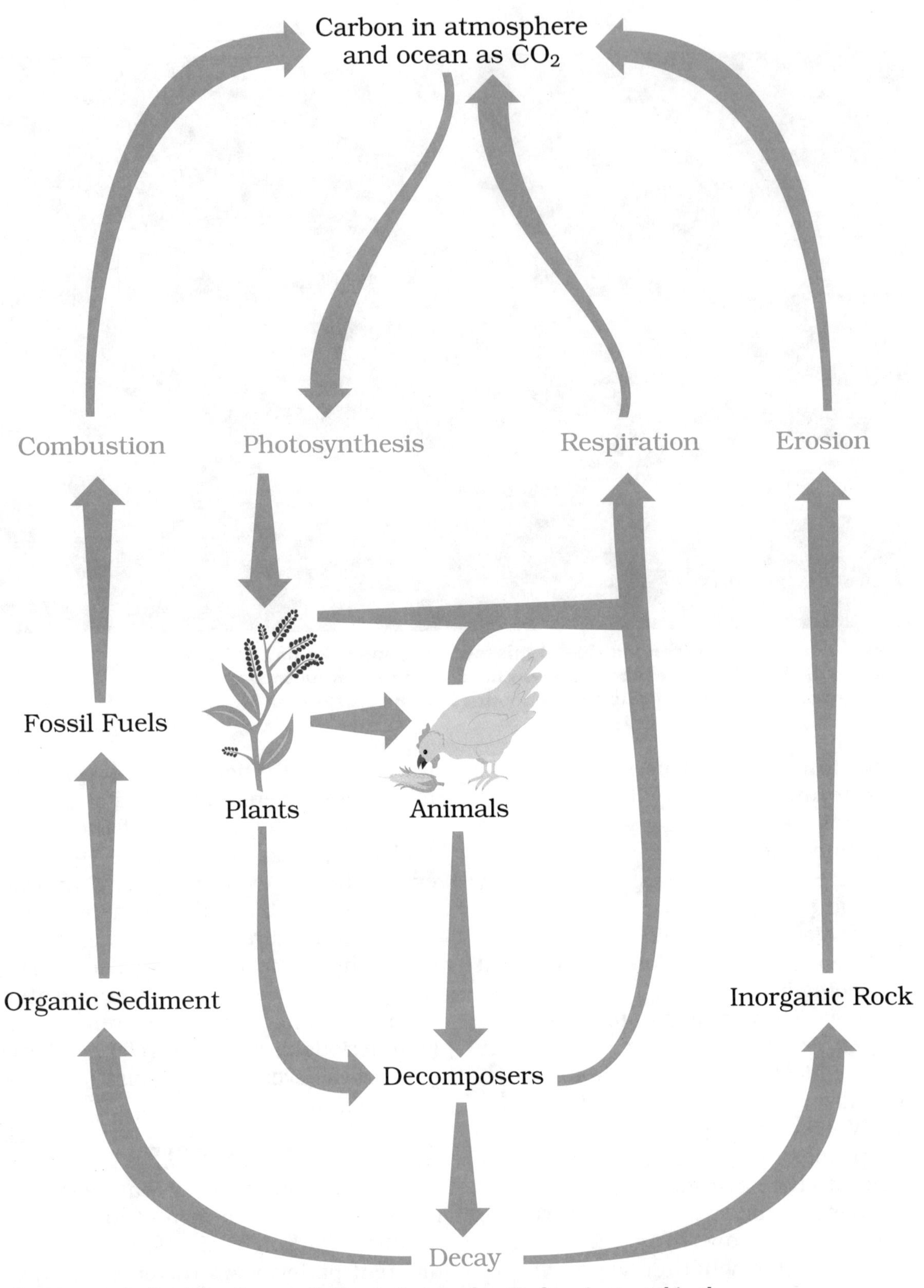

The arrows indicate pathways in the carbon cycle. Carbon is stored in the atmosphere, the ocean, rock layers, fossil fuels, and the bodies of plants, animals, and microbes.

The carbon in these shells may one day be part of a rock or an oil deposit.

plete a cycle may be long or short, depending upon the pathway.

One pathway is between living organisms and the atmosphere and oceans. In this pathway, carbon is taken up by producer organisms for photosynthesis. The producers, such as green plants, get the carbon from carbon dioxide. The atmosphere and oceans are great reservoirs of carbon dioxide. The producers use the carbon to build sugar and other needed chemicals. From the producer organisms, carbon moves through food webs to consumer and decomposer organisms. During the decay of dead organisms, carbon returns to the reservoirs in the form of carbon dioxide. Some carbon returns to the atmosphere and ocean directly from living organisms as a result of respiration, which releases carbon dioxide into the environment.

The recycling time for carbon moving from atmosphere to organisms and back can be brief or long. This depends upon how many trophic levels carbon moves through during its cycle. Look at the diagram on page 44. Consider a pathway in which carbon goes from air to plant and back via photosynthesis and respiration. This recycling may take only hours or even minutes.

Other pathways in the carbon cycle require much more time for recycling. One pathway transports carbon from organisms to rock. An example of this pathway occurs when shells of marine

By burning fossil fuels in motor vehicles, people are releasing vast amounts of carbon dioxide into the atmosphere.

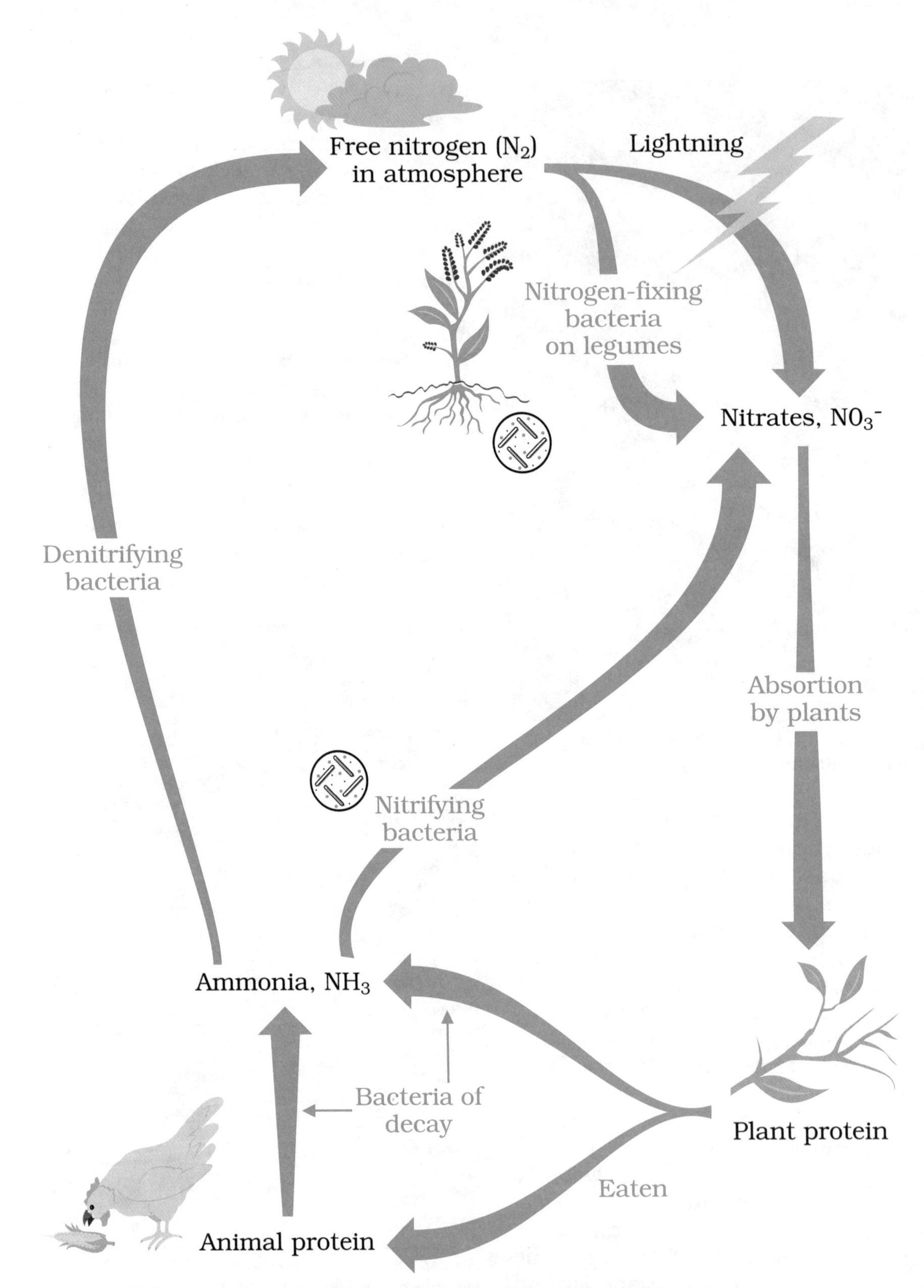

The arrows indicate pathways in the nitrogen cycle. Different kinds of bacteria in this cycle make nitrogen from the air available to plants, and recycle nitrogen from organisms back to the air.

animals are buried at the bottom of the ocean. These shells, which include carbon, eventually form rock, such as limestone. Much of this rock today was first formed at the bottom of ancient seas. Limestone in land ecosystems erodes when exposed to water and air. Erosion by water dissolves minerals with carbon. Eventually, some of that carbon is recycled to carbon dioxide.

Sometimes, the remains of dead organisms are buried and only partly decayed. This pathway leads to the formation of oil and natural gas, which are made up mainly of carbon and hydrogen. Examine the diagram on page 44. It takes millions of years for oil and gas to form.

This is a very significant pathway in urban ecosystems. Today, humans are recycling vast amounts of carbon by burning fossil fuels, which releases carbon dioxide into the atmosphere.

THE NITROGEN CYCLE

Nitrogen is the most abundant gas in the atmosphere. It makes up 80 percent of the air you breathe. This nitrogen reservoir is not used directly by most organisms. Unlike oxygen, nitrogen you breathe does not enter into chemical reactions, so it cannot be taken up into your cells. You must get your nitrogen from the food you eat.

How does nitrogen get from the atmosphere into the bodies of organisms? Bacteria play a starring role in this event. Certain kinds of bacteria in the soil are able to combine nitrogen from the air with oxygen to form **nitrate**. This is a process called nitrogen fixation. It means that the nitrogen becomes available for use by plants.

Some of the nitrogen-fixing bacteria live on the roots of plants such as clover, peas, soybeans, alfalfa, and their relatives. These are members of a group of plants called **legumes**. Land ecosystems need legumes to restore the supply of nitrogen in the soil.

Some kinds of nitrogen-fixing bacteria live on the roots of plants that are not legumes, such as alder trees. In addition, some aquatic bacteria can fix nitrogen for intake by organisms in aquatic ecosystems. It is interesting that lightning can also change nitrogen in the air to nitrate, as the diagram on p. 50 shows.

Nitrates in the soil are taken up by plants to make protein. If you keep a garden, you add nitrate to the soil when you add fertilizer. Like carbon, nitrogen can pass through the various trophic levels in ecosystems when organisms consume one another. Eventually, bacteria of decay break down plant and animal protein into ammonia. This is a chemical made of nitrogen and hydrogen. Ammonia in the environment is acted upon by another type of bacteria called denitrifying bacteria. These microbes release nitrogen gas back into the atmosphere. This event completes the cycle.

Lab Study

How is carbon dioxide cycled through an ecosystem via photosynthesis and respiration?

Materials

5 large test tubes
1 straw
1 bottle of bromthymol blue (BTB)
2 sprigs of *Elodea* plant
water
4 rubber stoppers to fit test tubes

Procedure

Part 1

1. Fill a test tube with water. Add three drops of BTB.
2. Note the color of the water.
3. Insert the straw into the test tube. Carefully exhale into the straw. Observe what happens to the color of the water.
4. BTB is an indicator that turns yellow or green in the presence of carbon dioxide. When the carbon dioxide is removed, it turns blue-green in color.

Part 2

1. Fill 4 test tubes with water. Add 3 drops of BTB to each test tube. Exhale into each test tube as in Step 3 above. At this point the water in all four test tubes should be yellow.
2. Label these test tubes A, B, C, and D.
3. Add an *Elodea* sprig to tubes A and C. Put a rubber stopper in each test tube so that they are airtight.
4. Place tubes A and B in direct sunlight for about 20 minutes. Record the color of the water in each tube at the end of this time period in the chart provided by your teacher.
5. Place test tubes C and D in a dark place for 24 hours. Record the color of the water in each tube at the end of this time period in the chart provided by your teacher.

TEST TUBE	OBSERVATIONS	
	INITIAL COLOR	FINAL COLOR
A		
B		
C		
D		

Conclusions

1. Explain the cause of the color changes that took place in the test tubes.
2. What is the purpose of test tubes B and D?

For Discussion

1. Explain your results in terms of the following summary equations:

carbon dioxide + water + solar energy —photosynthesis→ sugar + oxygen

oxygen + sugar —respiration→ carbon dioxide + water + energy

2. How is carbon dioxide cycled through an ecosystem by a plant?

Extension

Design and carry out an experiment similar to the above experiment to find out about the role of animals in the carbon cycle. You might use an animal such as a water snail.

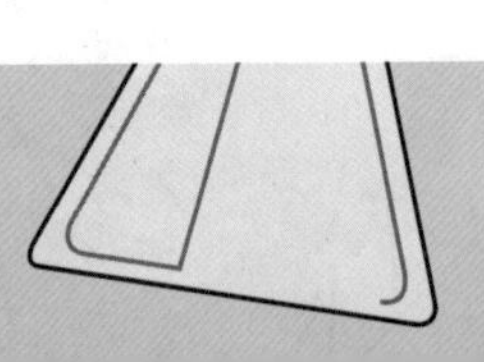

Field Study

As people become more environmentally conscious, there is increased interest in **organic** gardening and farming. At the base of the organic method of gardening and farming lies **composting**. Composting is the process by which garden and farm wastes are recycled naturally. Organic waste is broken down in the presence of oxygen and bacteria to form a rich humus end product that can be used to enrich and condition soil.

Many products are now available (in garden and hardware stores) that claim to start the composting process more quickly and efficiently. In this activity you are going to find out if these claims are justified.

Materials

2 large garbage bags
unsterilized topsoil
1 box of commercial compost maker
shovel
1 thermometer
grass clippings, leaves, other garden waste materials
water

Procedure

Part 1

1. Find a place in the schoolyard or your own backyard where you can place your compost bags so they will not be disturbed.
2. Place about 15 cm of topsoil in one of the garbage bags.
3. Add a 15-cm layer of grass clippings, leaves, and other garden waste materials.
4. Repeat Steps 2 and 3 until the bag is almost full, about 15 cm from the top.
5. Add just enough water to completely moisten the contents of the bag. The contents should not be soaking wet.
6. Label this bag "Soil." This will be your control.
7. Repeat Steps 2 through 5 substituting the commercial compost maker for the soil. Follow the package directions for how much to add to each layer.
8. Label this bag "Compost Maker."

Part 2

1. Examine the contents of the completed bags. Note the texture, color, and smell of the contents of each bag. Record your observations in the chart provided by your teacher.
2. Measure the starting height of the contents of each bag. Record these measurements in the chart.
3. Use a long laboratory thermometer to measure the temperature of the contents of each bag. Insert the thermometer into the center of the composting material. Record these measurements in the chart.
4. As the contents of the bag begin to dry out, add just enough water to keep it moist.
5. Tie the bags shut so that you trap as much air as possible inside them. Role the bags on their side to allow the air to mix with the contents. Then set the bags in a place where they won't be disturbed.

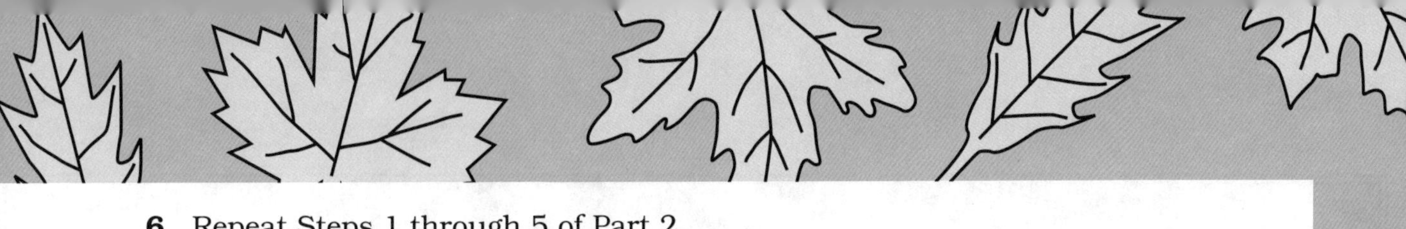

6. Repeat Steps 1 through 5 of Part 2 once a week for about 8 weeks.

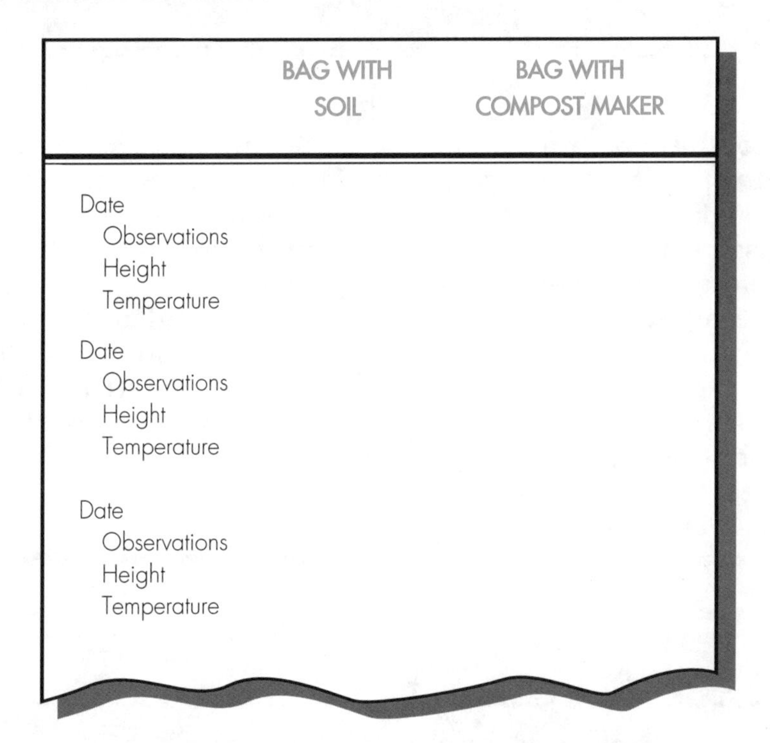

	BAG WITH SOIL	BAG WITH COMPOST MAKER
Date		
Observations		
Height		
Temperature		
Date		
Observations		
Height		
Temperature		
Date		
Observations		
Height		
Temperature		

Conclusions

1. Was there any difference in the rate of increase in temperature in the bags? What does this indicate?
2. Was there any noticeable difference in the rate at which the garden waste decomposed? Explain.

For Discussion

1. What is the source of the bacteria necessary to start decomposition of the garden waste in each of the bags?
2. What does the final stage of your compost look like?
3. Would you recommend a commercial compost maker based upon its performance in this activity? Why or why not?

Extension

Compare the growth of two identical plants, one planted in soil, the other in a mixture of soil and compost. Prepare two plant pots in the following manner. Mix equal amounts of soil and compost from this activity in one pot. The other pot should contain soil only. Plant flowers or vegetables in the pots. Compare the growth and development of the two plants for several weeks. Observe and measure the height of the plants and the growth of new leaves or flowers. Does the compost have an effect on plant growth?

You Solve It

The nitrogen cycle returns the nitrogen to the soil once it is removed by plant growth. Sometimes the nitrogen cycle does not replace the nitrogen in the soil fast enough to suit farmers. Therefore nitrogen-containing fertilizers are added to the soil to increase the growth rate and yield of crops.

1. Find out how much synthetic fertilizer is produced and used in the United States. What is the composition of most fertilizers?
2. There is a great deal of debate surrounding the use of synthetic fertilizers. Research the pros and cons of this issue. Present this information in the form of a debate.
3. Organic gardeners and farmers do not use synthetic fertilizers. How do they fertilize their soil? Discuss the merits of this approach.

CHAPTER REVIEW

1. How are the transpiration and evaporation pathways in the water cycle similar?
2. Choose a pathway from the water cycle and describe how it may be affected when a city grows in size and buildings take over the open land surrounding the city?
3. Is there a connection between driving a car and the carbon cycle? If so, what is it?
4. In which of the cycles described in this chapter do bacteria of decay play a role? Describe that role in as much detail as you can.
5. How might the cutting and burning of forests around the world affect the cycles described in this chapter?

FOR DISCUSSION

1. How does water that runs down the drain of a sink reenter the water cycle?
2. If bacteria in the soil were killed off by the actions of humans, what would be the effect upon the nitrogen cycle? How would this affect you?

Case Study

Burning Questions: Learning from the Yellowstone Forest Fires

The steep hill near Roaring Mountain in Wyoming looked like a charred wasteland. As Ann Rodman remembers it, the ground was a dark, charcoal black. Large craters marked the spots where the roots of centuries-old pine trees had proudly stood just two weeks earlier. Most visitors to this spot would have thought the world had ended, or at least that the area had been the scene of a terrible and deadly tragedy.

But Rodman, a scientist who studies soil, was no ordinary visitor. She surveyed this badly burned piece of Yellowstone National Park just two weeks after one of the worst forest fires in recorded history. And she calls the experience "the chance of a lifetime." Like many other scientists, Rodman has spent much of the past five years studying the effects of this huge fire. The research has helped us to understand more about the way a forest renews itself.

Scientists have used fires like the one at Yellowstone to learn more about the role fire plays in forests' cycles of change. You may be used to an annual cycle of seasons. Ecologists are learning that forests and other ecosystems go through much longer cycles with changes along the way. For some forests, fire is part of the cycle of changes.

Fires set by lightning are part of nature in arid places like Yellowstone. Scientists realize that more and more flammable, dead timber builds up on the ground as forests age, making fires even more likely. Up until 1972, officials at the National Park Service tried to stop all forest fires. Of course, they were not always successful. But nothing anyone could remember compared with the 1988 Yellowstone fires.

On just one day—August 20, 1988—more square miles of Yellowstone Park were burned in 24 hours than during any 10-year period since 1872. Winds tore through the park at 70 miles per hour. Flames soared 300 feet into the sky and ultimately claimed nearly one million acres—most of them forests.

In the five seasons since the fire raged, researchers like Rodman are still learning from the event. What they have found is helping us understand forest ecology. For instance, Rodman explains that even up until the Yellowstone fire many experts believed that forest fires burn the area's soil and "sterilize" it, removing all the roots and organisms. Rodman has proved that the opposite is true.

Even in a huge fire like the one at Yellowstone, she explains, the flames usually move so quickly in forested areas that the heat can not penetrate enough to do severe damage to the soil. "It may look like a bunch of dead trees," she says, "but much, much more is going on out there. In fact, all the ash makes the soil's nutrient levels go way up for the first year or two after a fire, speeding up the forest's rebirth."

Fire Opens Pine Cones

Rodman's colleague Roy Renkin has made similar discoveries about the regrowth of trees in areas burned by fire. Renkin has discovered, for example, that some tree varieties depend on fires for their successful propagation and renewal. Some of the pine cones dropped by lodgepole pines (a type of evergreen tree) open to release their seeds only after they have been burned. After the Yellowstone fires, Renkin and other scientists found as many as 150,000 lodgepole pine seeds per acre. Not only did the fire release the seeds, it also left them in an ideal habitat. Lodgepole-pine seedlings grow best in open, well-lit spaces with little competition.

Roy Renkin is especially interested in aspen trees, which often share a vast, interconnected root system. As Renkin explains, "A forest fire is one of the best things that can happen to aspens." Because the trees are really the shoots of the same underground organism, he says, the roots thrive with the chance to send forth new growth. "Where you had one aspen tree before a fire," Renkin explains, "you might have 500 new ones" after a fire.

According to Ann Rodman and Roy Renkin, fire is a natural and healthy process that keeps a forest's ecosystem going strong. But this view is hard for many park managers and members of the public to accept. After seeing vivid pictures on the television news, the public saw the Yellowstone fires as a terrible tragedy. How could such a catastrophe happen, many wondered? How could the people in charge let a fire like this get so out of control in a piece of land cherished for its natural beauty?

When Yellowstone was first established, Congress instructed the National Park Service to preserve the forests from injury. But forest fires challenge this mission. They raise a complicated question that tests our understanding of ecology. When lightning sets a forest on fire, should people try to put it out, or let it run its course, viewing it as part of a natural process?

Let It Burn?

Since 1972, the National Park Service follows a "Natural Fire Policy." The idea is that the service would let most natural fires burn themselves out. But it has been hard for park-service officials to stick to this policy. The 1988 fires put the policy to its most severe test ever. The officials in charge of Yellowstone Park found themselves in a very difficult position. Almost three million tourists visit Yellowstone every year. Nearby towns rely on the visitors for their economies, and no one wants to see a national resource like Yellowstone burned to a crisp.

Before the fires began to rage over large areas, the nation's top official in charge—the secretary of the interior—decided to fight back in spite of the natural fire policy. The government set up a fire-fighting effort that included army and marine soldiers and a fire-fighting team from Canada.

In fact, the official government reports written after the Yellowstone fire note that the fire fighting may have caused more long-lasting damage to the park than the fire itself. Trenches were dug to encircle the fire, and heavy equipment was driven through formerly untouched areas, making deep scars in the land.

But the official reports also question the costs of a natural-fire policy given the shrinking size of government wilderness lands. Large fires like those at Yellowstone in 1988 cause hardship for park visitors and local residents alike. No matter what the ecological consequences, shopkeepers and motel owners feared the loss of tourists who would not want to visit a burned area.

Because of these concerns, the government is now looking into replacing the natural-fire policy with one in which park rangers actually start "prescribed" fires regularly. The purpose of intentional fires is to lessen the amount of burnable fuel in the forest and prevent huge fires from raging out of control.

As Ann Rodman puts it, "We need to realize that we aren't just managing a big area with trees and grizzly bears; we're trying to protect a natural system with its own dynamic processes."

4 ECOSYSTEM CHANGES

This forest fire in Yellowstone signals a sudden, dramatic change. What follows a big fire — recovery of the ecosystem's plant and animal life — is gradual, but just as dramatic.

4.1 Change and Stability in Ecosystems

For years before the Yellowstone fires, park managers wanted to maintain what they thought was the park's stable ecosystem. They thought that the forest would be preserved best if it changed as little as possible from year to year. To help avoid any change, the managers worked very hard to prevent most fires.

However, scientists now realize that the Yellowstone ecosystem is not as unchanging as once thought. They have found that the ecosystem is a patchwork of tree stands of varying ages and stages of development. This is a result of small fires destroying different portions of the park about every 25 years. In preventing smaller fires, humans had removed a natural force of change that affected the ecosystem for most of its existence. The recent fire at Yellowstone has changed people's ideas about what is destructive to an ecosystem.

What causes ecosystems to change over time?

FORCES OF CHANGE

Fires, rather than being a force of destruction, are a necessary part of an ecosystem's natural cycle of change. Fire rids an ecosystem of decayed organisms and promotes new growth. The prairie grasslands of the midwestern United States, for example, have periodic fires that maintain the ecosystem. The fires kill nongrass plants while leaving the grasses' roots intact. This promotes new grass growth with little competition and provides fresh grazing material for the animals of the ecosystem.

All ecosystems are open to change. The events that cause change may be abiotic or biotic. Some events, such as fires or severe storms, radically change an ecosystem in a matter of weeks, days, or even hours. This occurred in Yellowstone. Other events are really gradual processes. They cause slow changes in ecosystems that may not be noticed for tens or hundreds of years.

Abiotic events include fire, volcanic eruptions, earthquakes, droughts, floods, and climate shifts. Biotic events include a population's tendency to grow and **disperse**, the introduction of an outside species, and disease.

The fire at Yellowstone is a stunning example of an abiotic force causing a sudden and drastic change in an ecosystem. Where acres of tall lodgepole pines once stood, today there are huge fields of grasses, wildflowers, and young pine saplings. These species were able to grow because of the lack of competition from the mature trees.

Another example of sudden abiotic change occurred on the slopes of Mt. St. Helens, an active volcano in

Several years after the eruption of Mt. St. Helens, trees are growing back on this hillside.

As climate warms, glaciers retreat, leaving behind bare earth.

Washington state. In 1980, a huge eruption of ash and gases knocked down acres of trees and covered the surrounding area with ash. The forest ecosystem was destroyed in a matter of minutes. Within months, however, plant and animal life began to be reestablished.

Now the area is covered with flowers, shrubs, and small trees, with a return of most of the animal life.

Recall that all species have tolerance ranges that if surpassed can result in a serious reduction in population. A change in climate, for example, could be as subtle as a drop of 1°C over a century. However, that could be enough to surpass the tolerance range of a certain plant species that would then die off in that ecosystem.

Gradual biotic change occurs in an ecosystem as species reproduce. At times, certain conditions within an ecosystem can change, favoring one species over another. The population of the favored species would slowly increase. Over time, the increasing population of a species could outcompete—and thus eliminate—another species in the ecosystem.

For example, about 20,000 years ago much of North America was covered by huge ice sheets and the climate was significantly colder than it is today. However, by 7,000 years ago much of the ice had melted and the temperature had increased. Many species were able to move farther north into areas that had been outside their tolerance range during the colder period.

The earlier climate had favored species of trees, such as fir and spruce, that were adapted to survive cold temperatures and a moist environment. As the temperatures increased and the environment became drier, trees that thrive in warmer—such as pine, oak and maple—climates were able to compete for space with cold-climate trees. The pines, oaks, and maples were able to reproduce more quickly than fir and spruces and eventually replaced them in much of what is now the northern and eastern United States.

FORCES OF STABILITY

You might think that with so many forces working to change ecosystems it would be nearly impossible for any system to remain unchanged. However, some ecosystems contain the same combination of species for centuries and even tens of thousands of years. Such ecosystems often are referred to as stable. This stability is

Gradual change in an ecosystem: trees and shrubs (foreground) are growing on land that used to be covered by the glacier (background).

possible because of two elements in the ecosystem that work to oppose change. These elements are: (1) a stable climate and (2) limiting factors of the species.

Climate is of great importance to maintaining a stable ecosystem. Over time, species become adapted to the specific conditions of a climate. As long as those conditions continue, the species thrive. Plants are dependent on receiving the same amounts of rainfall at the same season each year. Most animals have a limited range of temperatures they can tolerate. For example, trees in a tropical rain forest are dependent on the constant and plentiful rain they receive year-round, while tropical birds could not survive without the warm temperatures.

The stability of ecosystems is also encouraged by limiting factors, such as available sunlight, amount of space, precipitation, food supply, and soil nutrients. These limiting factors provide resistance to one major factor of change—population growth. By limiting population to levels that the ecosystem can sustain, the ecosystems are kept stable.

SECTION REVIEW

1. Why would fire be considered a limiting factor that keeps a grassland ecosystem stable?
2. When a river floods its banks, it deposits a layer of mud and silt there. How might a flood cause a change in a surrounding marsh? What would happen to crop land along its banks?
3. What activities do you engage in that could change an ecosystem? Which activities of yours might keep the ecosystem stable?

Lab Study

How do ashes left behind after a forest fire affect the soil?

Is a big fire an ecosystem disaster? Or is the fire a part of the ecosystem's natural cycle of changes? Perhaps, after a fire is over, there are some benefits that can be identified and measured.

In this activity, you will examine the effect of ashes on soil quality.

Materials

pH paper
2 stirrers
warm water
2 thermometers
soil
4 plastic cups
wood ashes

Procedure

Part 1

1. Place an equal amount of soil in 2 plastic cups.
2. Put a 1-cm layer of wood ashes onto the soil in one cup. Label this cup A. Label the other cup B.
3. Insert a thermometer into the soil in each cup. The thermometer bulbs should be placed at the same depth in both cups, about 1 cm deep.
4. Set both cups in a sunny place. Record the temperature of the soil at the beginning and every 10 minutes for about 40 minutes. Record the temperatures in the chart provided by your teacher.

TIME	TEMPERATURE	
(minutes)	Cup A	Cup B
0		
10		
20		
30		
40		

Part 2

1. Mix some soil with wood ashes. The ratio of ashes to soil should be about 1 to 6.
2. Fill one cup about 3/4 full with the ash/soil mixture. Place an equal amount of plain soil in another cup. Label the cups according to their contents.
3. Fill each cup with warm water. Use a stirrer to mix the soil in each cup. Let the soil settle for about 10 minutes.
4. Place a strip of pH paper into the water in the ash/soil cup. Remove the strip and hold it up to the scale on the pH vial. Match the color of the strip to the colors on the pH scale. Record the pH on the chart provided by your teacher.
5. Repeat Step 4 with the cup of plain soil.

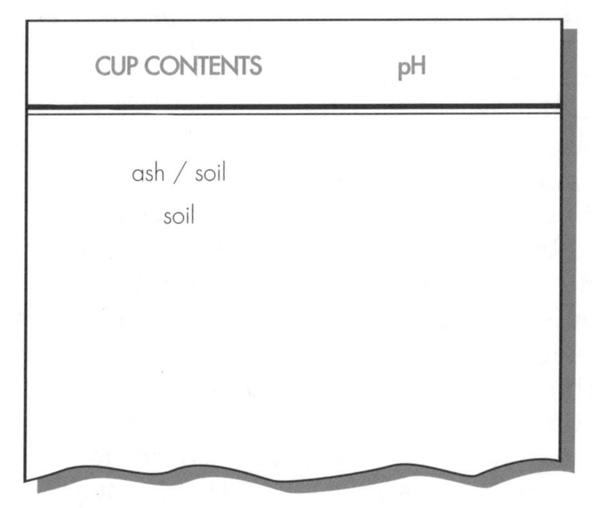

CUP CONTENTS	pH
ash / soil	
soil	

Conclusions

1. How was the temperature of the soil affected by the layer of wood ashes?
2. How did the ashes affect the pH of the soil?
3. How might the presence of wood ashes affect the germination of seeds in an area cleared by forest fire?

For Discussion

1. Research the affect of soil pH on the growth of various plants. Use this information to explain how the presence of ashes might affect the ecological succession in an area cleared by a forest fire.
2. Find out how pH affects the availability of soil nutrients. Would the presence of wood ashes have a positive or negative effect on the mineral content of the soil?

Extension

Mix some dirt with wood ashes. The ratio of ashes to dirt should be about 1 to 6. Place this mixture in a pot. Fill another pot with plain dirt. Plant coleus seeds in both pots according to package directions. Compare the growth and development of the coleus plants for about 4 weeks. How might you account for any difference in development?

4.2 Ecological Succession in Ecosystems

When an ecosystem has been destroyed, it can only be replaced through a process called **ecological succession**. Ecological succession is the process by which species of an ecosystem are gradually replaced by other species. This replacement is a result of competition between differently adapted species.

Succession proceeds through a number of stages with significantly different species populating the ecosystem in each stage. Remember, however, that at any stage of succession an event may occur that can set the entire ecosystem back to an earlier stage. Then the process must begin again.

The pocket gopher can be a pioneer species in a changing ecosystem.

How do new species repopulate destroyed ecosystems?

PIONEER POPULATIONS

An ecosystem that has been wiped out by a catastrophe is first repopulated by **pioneer species**. Pioneer species have the ability to reproduce and spread out quickly. In Yellowstone, a pioneer species is the pocket gopher seen below. Why would the adaptations of pioneer species be important for establishing a population in such an environment?

The pioneer community of species prepares the environment for other species by breaking down rock to form soil, adding nutrients to existing soil, and providing food sources. Then it is possible for other species to invade the ecosystem and compete with the original species for resources. If the new species are successful, they inhabit the ecosystem until they are replaced with still other competing species. Each stage of succession is the result of competition—a continuing struggle among various species for a place in the evolving ecosystem.

What type of change occurs as succession progresses?

STAGES OF SUCCESSION

While there is no set sequence for which species may follow another in any one ecosystem, the stages of succession do follow a trend. The faster growing species present at the earliest stages are replaced by species that tend to grow more slowly, have larger individual size, and live longer. For example, the fireweed plants that grew weeks after the Yellowstone fire were joined within a year by small shrubs and later by lodgepole pine saplings.

All of these changes in the characteristics of species are a result of the difference in reproductive strategies between the earlier and later species. In the earlier stages, species that grow and disperse rapidly can best exploit the unused resources and are thus favored. However, as more species join the ecosystem, there is increased competition for resources. Species that can best obtain those resources are then favored. Larger organisms tend to grow more slowly, and therefore do not reproduce until a later age than faster-growing species. Larger species also tend to have longer life spans. Thus one individual is able to reproduce repeatedly during its lifetime. This gives the slower growing species an advantage in competing for resources.

Another significant change that often occurs as succession progresses is the increase in the number of species that populate the ecosystem. Pioneer communities consist of very few species, so ecologists say that they have low **species diversity**. This results in a simple food chain of mostly producers and a few herbivores. These species need few nutrients and use most of their energy to reproduce. With each progressive stage, the diversity of species increases. The food web thus becomes more complex.

Under what conditions does succession end?

CLIMAX COMMUNITIES

The kinds of plants and animals present at the early stages of succession are not as important to development of the ecosystem at later stages. The diagram on p. 64 shows the changes in vegetation that begin with an abandoned field. Notice that the species present at the earlier stages slowly give way to other species. Those species with wider tolerance ranges exist within the ecosystem for a longer period of time.

Succession will continue until the ecosystem sustains a **climax community**. A climax community exists in an ecosystem when no other combination of species is able to successfully compete with or replace that community. The climax community maintains a stable ecosystem that will change little if undisturbed. The lodgepole pine forest of Yellowstone is a climax community.

Another climax community can be found in the region of central North Carolina. When European settlers arrived in this area hundreds of years ago, they cleared away a climax community of oak and hickory trees and planted crops in its place. Decades later some of the farmland was abandoned and left bare. Over the following 150 years, a series of successional communities occupied the land.

At first, the soil of the abandoned farmland was taken over by quick-growing annual weeds. Though less favored, grasses and small perennial plants were able to get a foothold among the annuals. Perennials are plants that survive from season to season. The roots of grasses also survive from season to season allowing them to resprout each spring.

Because of their longer life spans, the grasses and perennial plants were able to outcompete the annuals and dominated the next stage of succession in the changing North Carolina ecosystem. Once the soil was prepared to

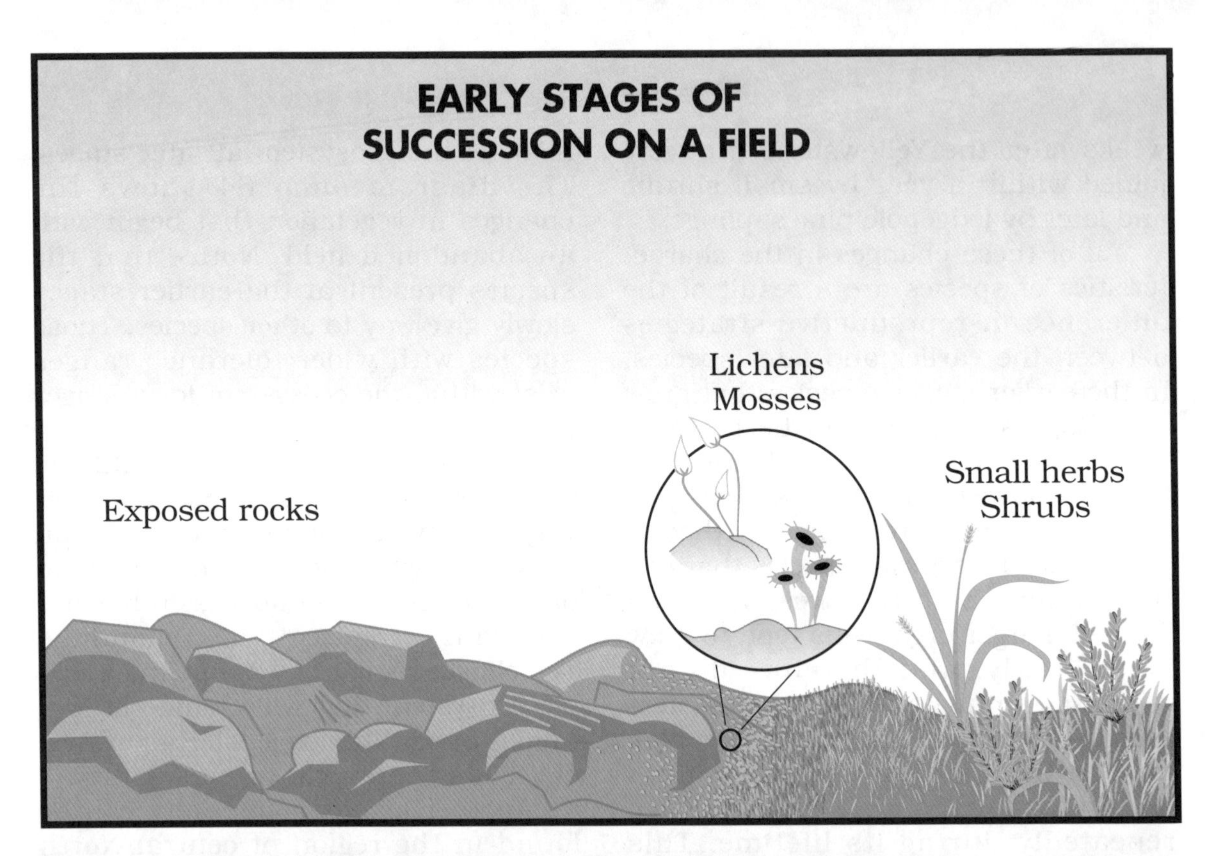

This model of ecological succession shows changes that occur on one site over a span of 100 years or more. Pioneer plants begin the process, and are gradually replaced by trees.

support small perennial plants, larger plants, such as shrubs, were able to take root. Their larger size allowed the shrubs to outcompete the smaller plants by using up more of the available resources. Thus, the ecosystem became dominated by shrubs.

The shrub stage gave way to a pine-forest stage. The pine trees were able to take over the ecosystem for two reasons. First, as the pines grew taller and taller, they were able to receive more sunlight while shading the shrubs. Second, they were able to disperse their seeds farther. In time, the pine forest was replaced by oak and hickory trees for the same two reasons. The forest of oak and hickory trees became the last stage of the succession—the climax community—and cover much of central North Carolina today. Those trees were able to successfully outcompete and exclude any other combination of species.

Common sense might tell you that a climax community, because it is stable, will survive a stress more readily than a less mature ecosystem. In one respect this is true. In an ecosystem with low diversity, the system is more vulnerable to a slight change in the ecosystem. All of the species are dependent on a small range of conditions and could be destroyed if any of those conditions are disturbed. For example, a field of grass will less likely survive a two-year-long drought than would a forest of 200-year-old oak trees.

However, after an event such as a 5-year-long drought, or a volcanic eruption that destroys an ecosystem, it is the younger community that can more quickly regenerate once those stressful conditions have ceased. Unlike the pioneer communities, climax communities use little of their overall resources to reproduce. Most of their energy is expended in maintaining the life functions of the individuals in the community. The mature ecosystem takes the longest to be restored to its predisaster state because of the slow reproduction, growth, and dispersal rates of the organisms within it. The entire sequence of succession must be repeated before the climax community can be reestablished. This can take hundreds or even thousands of years.

SECTION REVIEW

1. What examples of succession can you think of that can be observed near your home?
2. How could a species of grass change an ecosystem so as to cause its own elimination from that ecosystem?
3. Why is the ecosystem of a planted cornfield more likely to be destroyed by disease than a forest would be? What types of stress would the forest be more vulnerable to?
4. Succession on abandoned farmland in central North Carolina led back to the original oak and hickory forest. What can you conclude about the climate of this region?

Field Study

A pine forest, a salt marsh, or a tropical rain forest does not spring forth full blown from the ground. Each of these ecosystems develops over years or even decades. The process begins when a simple community of pioneer species takes over a patch of cleared earth. These early species are subsequently joined or replaced by other species to form a new community. The gradual, sequential replacement of species in an area is called ecological succession.

In this activity you will clear a small patch of earth and observe the early stages of ecological succession.

Materials

journal
hand lens
spade or shovel
rake
meter stick
string
4 stakes

Procedure

1. Select a small area near your home or school that could be cleared and observed over a 2-month period.
2. Measure off a 1-meter-square portion of the area you select. Use the stakes and string to mark this plot of land.
3. Pull up any weeds, small plants, or saplings that are growing in the plot.
4. Use the spade or shovel to turn the soil over so that only dirt is visible in the plot.
5. Use the rake to break up any large clumps of dirt and then smooth out the soil.
6. Draw a map of your plot and the area around it. Include the locations of trees or woods, large rocks, streams, buildings, parking lots, sidewalks, etc.
7. Observe your plot every week for about 2 months. Record the types and estimated number of each species that you find growing in the plot each week.

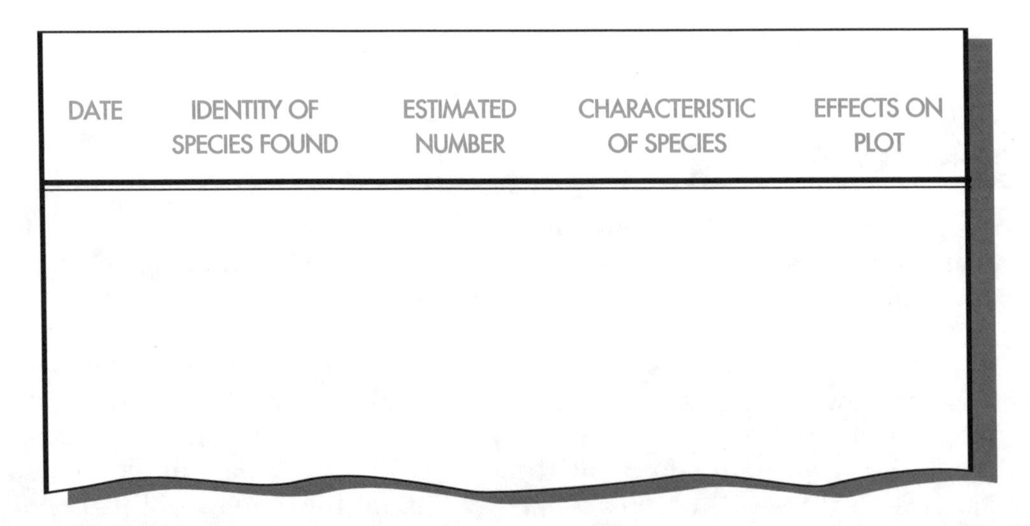

DATE	IDENTITY OF SPECIES FOUND	ESTIMATED NUMBER	CHARACTERISTIC OF SPECIES	EFFECTS ON PLOT

8. As each new species appears in your plot, describe its characteristics and its effect on the plot.

Conclusions

1. Identify the pioneer species that colonized your plot.
2. Identify the biotic and abiotic factors that influenced the colonization of your plot.
3. How does your plot compare to the area surrounding it? Would you classify your plot as a mature or immature ecosystem? Why?
4. Over the 2-month period, did the pioneer community look as if it were being replaced by an intermediate community? If so, identify species that make up the intermediate community.

For Discussion

1. Can the type of disruption that you caused when you cleared the plot of land for this investigation be caused by any natural causes? Any other human activity? Explain your answer.
2. Consider the following statement: Farmers keep an ecosystem at an early stage of succession. Do you think that this is true? Why or why not?

Extensions

1. Can you predict what things might follow the pioneer species on your plot if it were left alone for a year? A decade? Explain your predictions.
2. Study the maps, observations, and conclusions of your classmates. How do their results compare? How can you account for any differences in the sequence of appearances of species, the types of species, and the number of species that colonized the plots?
3. Write a report on the succession that occurred after the volcanic eruption of Mt. St. Helens in Washington state. What were the pioneer species? What types of plants may yet colonize this area?

A small part of a forest ecosystem can recover after being cut. When most of the ecosystem is damaged, however, much of the wildlife can be lost.

4.3 Humans Change Ecosystems

Up to this point, you have learned of all the ways in which nature changes itself. Unfortunately, this does not take into account the one factor that is responsible for most of the changes to ecosystems occurring today—*humans*. The list of ways in which people alter the environment is long but includes pollution, settlement, farming, grazing livestock, mining, irrigation, overhunting, pest control, deforestation, altering waterways, introduction of outside species, and fire. All of these actions, and many others, can drastically change an ecosystem.

How do people damage stable ecosystems?

PEOPLE DISRUPT ECOSYSTEMS

Of the alterations that humans make to the environment, the ecosystems most harmed are the climax communities. These, as you have read, are the least likely to survive a damaging effect and may require 100 to 1000 years to regenerate. If environmental conditions change dramatically, ecosystems may recover, but some species of wildlife may die out. The loss of wildlife changes the biotic interactions in ecosystems.

Examine the photo on p. 68. This illustrates the practice of cutting down and then burning sections of forest to clear it for farmland, pasture, or building of roads and towns. The ash of the burned wood temporarily enriches the soil, allowing productive farming for a few years. However, the nutrients soon are used up, the field is abandoned and, without the previous forest cover, the soil washes away. The conditions are such that it is no longer possible for the rain forest to regenerate over that area even after the land has been abandoned.

You Solve It

The primary goal in managing Yellowstone Park and other national parks is to maintain their ecosystems as near to a natural condition as possible. Since ecosystems change, plant and animal populations will change as well. Protecting national parks, therefore, means allowing natural cycles and processes to occur while allowing people to visit the parks.

Natural processes include plant succession, fluctuations in relative abundance of animals, and a wide variety of natural disturbances, such as forest fires, droughts, and floods. Of these natural processes, fire is one over which park managers have tried to exercise some control. Very sophisticated approaches to controlling and extinguishing forest fires have been developed.

Up until 1972 extinguishing natural forest fires was a component of wildlife management. But recent research has pointed out that this approach may be a major departure from natural conditions. As a result, there has been a movement to allow natural fires to run their course in an effort to replicate natural cycles that occur in healthy ecosystems.

1. Identify the advantages and disadvantages of natural forest fires to the ecosystems where they occur. What evidence is there that fire extinguishing interferes with natural processes?
2. Research the Natural Fire Plan that has been in effect in Yellowstone Park since 1972. Identify the criteria used to determine what type of fires will be permitted to run their course in the park. Explain the success or failure of this plan in maintaining the "pristine condition" of the park.
3. There is currently a great deal of debate about the role of fire in the management of our national parks. The political, social, and ecological implications of a natural-fire policy are often in opposition. What role do you think fires should play in wildlife management? Explain your answer and offer supporting evidence for your stand.

SECTION REVIEW

1. Recent U.S. government policy on wetlands has been one of "no net loss." This means if a developer wants to destroy 1 acre of wetlands, 1 acre of wetlands must be created or restored in its place. What are the benefits of and problems with this policy?
2. Why is preserving existing ecosystems preferable to attempting to repair them after humans have damaged them?

FOR DISCUSSION

What do you think is the best response to forest fires in our national parks? How would your view be different if you were the manager of a motel in a town next to Yellowstone Park? Or if you were the secretary of the interior?

5 HUMAN VALUES AND ECOSYTEMS

What should people do with the beautiful shoreline shown above? Some people may wish to keep it as it is; others may prefer to build homes or hotels.

All of us make decisions every day based upon our values. Our values define what is important to us. These might include religious beliefs, good performance in school, or the relationships with our family and friends. Our values tell us what is right and wrong. They help us to decide how to behave with regard to the world around us.

Not everyone shares the same set of values. When values clash, people may have more difficulty solving problems together than when values are shared.

5.1 Values and Environment

Conflicts in values affect people's ability to solve environmental problems, too. When a group of persons are trying to decide how best to protect the environment, some persons will have different opinions about the proper actions. These different opinions will be based upon their different values.

THE TRAGEDY OF THE COMMONS

In today's world, conflicts arise when people debate environmental issues. One such conflict is between short-term benefits to individuals versus the long-term benefits to a group or community.

Garrett Hardin, an ecologist, illustrated this conflict with an essay called "The Tragedy of the Commons." Earlier in history, towns and villages had grazing land that belonged to every resident. This grazing land was called the commons. Everyone in the town could graze their livestock on the commons. It was a resource owned by no single individual, but shared by all.

People must decide what to do with this land when all the coal has been mined.

An individual livestock owner, of course, would want to have as many animals feeding on the free grass as possible. The reason for this is clear. The individual animal owner benefits by raising livestock that can be used for food or sold in the market. The individual, however, pays nothing to support or care for the grass on the commons.

When every livestock owner does this, though, the grass is likely to be overgrazed. Each owner might ask, "What harm could one more animal do?" Most individuals would answer, "very little harm." It's easy for each animal owner to understand the benefit of putting more livestock on the commons. It's difficult to recognize the costs of doing so. The cost of overgrazing is to degrade the commons. As the grass is worn away, fewer and fewer animals can graze there. Then all livestock owners are faced with reducing their herds.

This is Hardin's lesson about the "tragedy of the commons." When all individuals in town try to get the greatest individual benefit from the commons, everyone stands to lose the valuable resource through overuse.

Our rivers, forests, air, soil, and oceans are our commons. We are beginning to see that we need to view the entire biosphere as a commons. The nations of the world are sometimes in conflict over the global commons. At the Earth Summit in Brazil in 1992, the nations of the world struggled with how to deal with concerns like global warming and extinction of species. When everyone tries to make as much money as possible from the commons, the resources can be degraded and even destroyed. We have conflicts over how to use the commons because we all have different values.

ENVIRONMENTAL ETHICS

Throughout the world, different cultures and individuals have their own environmental ethics. Environmental ethics are the ways in which persons believe they should treat the environment. Environmental ethics are standards of what behavior is good for ecosystems and what behavior is destructive. These ethics are shaped by values, by what persons think is important. Environmental ethics can be divided into two basic categories.

A human - centered ethic Some people believe that the earth's resources (the abiotic and biotic components of the earth's ecosystems) are here for us to use in any way that we wish. These people believe that we are the dominant species on earth and that everything is here for our benefit. We could call this the human-centered ethic.

An earth-centered ethic The other view could be called the biocentric or earth-centered ethic. People with earth-centered ethics believe that the entire biosphere and all of its individual parts are valuable. These people feel that in order for humans to be healthy, the earth must be healthy. The entire biosphere must be cared for.

According to the biocentric ethic, nature is valuable whether it can be used by humans or not. This ethic sees all members of an ecosystem as belonging to a community in which resources are limited.

Aldo Leopold Aldo Leopold worked for the U.S. Forestry Service during the first half of this century. He is considered to be the father of environmental ethics. He wrote a book entitled *Sand County Almanac*, published in 1949. Regarding changes that we may wish to make in the environment, Leopold wrote that "a thing is right when it tends to preserve the integrity, stability, and beauty of the biotic community. It is wrong when it tends otherwise."

This land used to be a strip mine. What could it become next?

ENVIRONMENTAL ETHICS IN HUMAN SOCIETIES

Biocentric values can be found in many of the world's religious teachings, including the Native American

This marsh is a commons. Do you think that people can use it without destroying it?

cultures. People who live close to the land and depend on the land for their survival often have a strong respect for the environment. Many people who are presently concerned about the environment believe we can learn from Native American cultures and from other groups of people who live close to nature.

People who speak and write about an environmental ethic are concerned with building a sustainable society. This means that they believe humans cannot continue to deplete the world's resources, both biotic and abiotic, at the current rate. A sustainable society is one that protects the earth's ecosystems while trying to improve the standard of living for future generations.

This is not a new idea. In their mythology, the Iroquois tell of caring for the world for the seventh generation. They believe they must treat the earth in a way that will provide people seven generations in the future with the same environmental quality.

SECTION REVIEW

1. List some of your values. Focus on those values that you think affect the environment. Then write the ways in which you think your values affect your local environment.
2. Try to think of an example in your own community that is like the story of the "Tragedy of the Commons." Explain the situation.
3. Which environmental ethic do you believe in—biocentric, human-centered, or somewhere in between? Explain.
4. For some people, land is a possession. For others it is a natural resource to be shared. Is there a difference between these two ideas? If so, what is the difference?

5.2 Restoring Ecosystems

After decades of destroying ecosystems, humans are now realizing that they can, and must, begin to repair some of the destruction. People in the United States and around the world are now working to repair ecosystems by restoring forest, wetland, and grassland habitats.

The original Kissimmee River was a winding stream surrounded by wetlands.

One example is an effort to restore portions of the Kissimmee River ecosystem in Florida. Between the 1960s and 1970s, the U. S. Army Corps Of Engineers changed the winding, 103-mile long, slow moving river into a straight, 56-mile long, fast-moving, canal. This was done to control local flooding. The Corps drained twenty-six thousand acres of wetlands to straighten the Kissimmee. The river's abundant fish died from lack of oxygen. The water fowl such as egrets, herons, and migrating ducks all soon disappeared from lack of habitat and food sources.

Today the Corps of Engineers is restoring some of the Kissimmee River's original winding course and 43,000 acres of wetlands ecosystem. An experimental section already has been repaired with apparent success. Many of the native plants and some bird species such as egrets and herons have already returned to the restored section. One of its goals will be to reduce chemicals that drain through the canal into Lake Okeechobee, the second largest fresh water lake in the nation. It will also help restore the fresh water supply to southern Florida's cities.

Restoration of ecosystems means working with both biotic and abiotic factors. Scientists in the northern United States have worked to repair lakes damaged by acid rain. They add lime to the lakes to neutralize some of the acid. Changing water quality alone won't restore all the living organisms, however. It is often necessary to bring in new plant populations from other places. It's also necessary to reintroduce animals, such as frogs or fish, that were wiped out by the acid rain.

Restoring damaged ecosystems is a good idea, because all people benefit from the things ecosystems provide us: healthy soil, water, air, recycling of materials, and natural beauty. Scientists tell us that it may be impossible, however, to restore all the interactions among living organisms and abiotic parts of an ecosystem. For this reason, people should focus on conserving the natural ecosystems we still have.

Field Study

It is common to think of your village, town, or city as a community. Animals and plants live in communities too. Populations of various plants and animals living and interacting in a given area are called a natural community. A natural community can be as large as a pine forest or as small as a fallen log. Communities can be found near a city sidewalk or in a country pond.

No matter what size or type of community you study, the same procedure is followed. First you set the boundaries of the community. Then you identify all the organisms found within the boundaries. And finally, you discuss the interactions among the organisms found there. In this activity you will follow this procedure and study a community near your school or home.

Materials

journal
markers
ruler
reference books and field guides

Procedure

Part 1

1. Select an area near your school or home. The size of the area you select will be dependent upon the time available to thoroughly explore it. Your teacher will give you some guidelines about the size of the area that you should select.
2. Make a map of the area you selected. Include notable objects that border it. For example, if you select a vacant field in the middle of a city block, include nearby buildings, parking lots, roads, streets, etc. If you select a fallen tree, include standing trees near by and large rocks or other objects.
3. Begin to explore your area to find out what types of organisms live there. Use reference books and field guides to help you identify the organisms you find.
4. List the organisms in column 1 on the chart provided by your teacher. Estimate the numbers of organisms in each population and record this number in column 2 of the chart.

Part 2

1. A community consists of the interactions between and among the populations of organisms that live there.
2. Look at the organisms that you listed in column 1 of the chart. What types of interactions did you observe between or among the organisms your found there? Describe these interactions in the last column of the chart.
3. Can you infer any interactions among the living things you found in your area? If so, include these in the chart also.

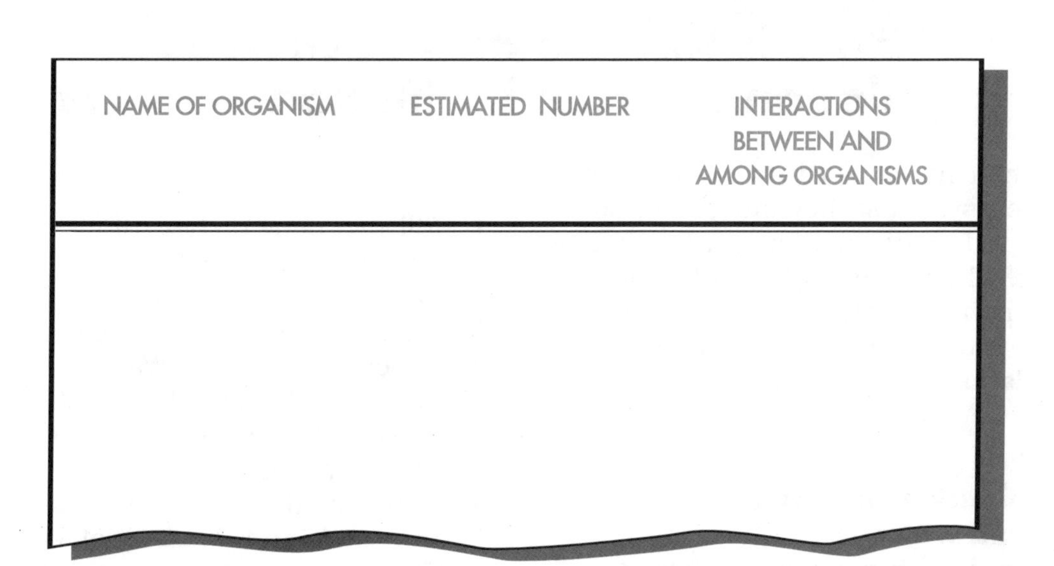

NAME OF ORGANISM	ESTIMATED NUMBER	INTERACTIONS BETWEEN AND AMONG ORGANISMS

Conclusion

1. Are there any dominant organisms in your area?
2. Is the area you selected a natural community? Why or why not?

For Discussion

1. What organisms found in your area depend upon other organisms for survival? Explain.
2. Do you think a large or a small community is more likely to change in response to a change in the size of a given population? Explain.

Extension

A community interacting with the abiotic parts of the environment is called an ecosystem. Take another look at your community. What interactions can you identify among the organisms you identified and between these organisms and their physical environment? Discuss these interactions in detail.

Lab Study

How is the size of a duckweed population affected by human activity?

When humans try to restore damaged ecosystems, they must have some knowledge of how organisms are affected by the damage. In this activity you will investigate the impact that damage to water quality can have on a population of aquatic plants.

Materials

25 duckweed plants
pond water
1 pair of tweezers or forceps
5 petri dishes
1 box of salt
1 bottle of vinegar
1 bottle of plant food
1 medicine dropper
1 balance

Procedure

1. Fill five petri dishes almost to the top with pond water.
2. Label one dish "Control." To this dish add 5 duckweed plants.
3. Label a dish "Salt Water." To this dish add 2 g of salt and stir until dissolved. Then place 5 duckweed plants in this dish.
4. Label a dish "Acidic Water." To this dish add 7 drops of vinegar. Then place 5 duckweed plants in this dish.
5. Label a dish "Plant food." Follow package directions and to this dish add the appropriate amount of plant food. Then place 5 duckweed plants in this dish.

DAY	NUMBER OF PLANTS				
	CONTROL	SALT WATER	ACIDIC WATER	PLANT FOOD	DARK
1	5	5	5	5	5
3					
5					
7					
ETC.					

6. Label the last dish "Dark." Place 5 duckweed plants in this dish.
7. Cover all the dishes. Place the dish labeled "Dark" in a place where it will get no light. Place the other dishes in a sunny window.
8. Every other day for 3 weeks count the number of duckweed plants in each dish. Record the number in the chart provided by your teacher.
9. As water evaporates from the dishes, add additional pond water.

Conclusions

1. Graph the number of duckweed plants versus the day you counted them. Use a differently colored line for each environmental condition.
2. What conditions promote duckweed growth? What conditions inhibit duckweed growth? Are there any conditions that do not appear to affect duckweed growth? Explain.

For Discussion

1. How might each of the conditions that you created in the petri dishes be created in real ponds as a result of human actions?
2. What biotic parts of a duckweed's ecosystem might affect its size?

Extension

What actions can you take to restore the duckweed population? How might you improve water quality in the salt water and acid water environments?

There are other abiotic parts of the ecosystem that might affect the growth of a duckweed population. Brainstorm ideas with other students and then design and carry out experiments to test your ideas.

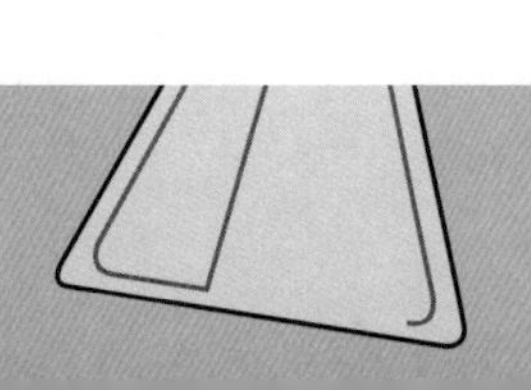

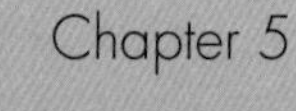

SECTION REVIEW

1. Could duckweed plants be used as a test organism to see how well restoration of a damaged lake ecosystem is working? Explain your answer.
2. Damage from flooding can be measured in dollars. Thus the value of flood control in the Kissimmee ecosystem can be estimated in dollars. Can you think of a way to estimate the value of an undisturbed wetland-and-river ecosystem?

FOR DISCUSSION

Much like the fire prevention policy in Yellowstone, the Kissimmee River restoration has its supporters and opponents. How would you react if you were a farmer whose lands would be flooded as a result? How would you react if you owned a fishing and boating equipment store in the Kissimmee ecosystem?

Youth Environmental Action

"You haven't been in mud until you've been up to your knees in it," Dave Foertsch said.

"It's tedious and hard, but that's what we're here for," said Becky Burgess. "I'm so excited because this is what I want to do: something in the environment."

Dave and Becky were talking about planting tens of thousands of individual plants of salt marsh grass. They and 23 other high school students spent a week on Wye Island in the Chesapeake Bay, Maryland. By the end of the week, the students had planted grass along one-third of a mile of shoreline.

Before the students could do the outdoor work, they had to spend months planning their project. They had to solve such problems as where to get fresh water for a stay on an island with no buildings, no running water, no toilets, and no showers. They planned meals, made up grocery lists, shopped for bargains on food, and decided what cooking equipment to bring. They rented portable toilets, made up lists of personal equipment each person had to bring, and set up rules for their island stay. They prepared the daily work schedule and divided themselves up into work groups.

Then, in June 1992, the students camped out on Wye Island for a week, supervised by Dr. Patricia Neidhardt, science teacher at Broadneck Senior High School in Annapolis. Most of their daily work was spent planting two kinds of grass called Spartina. Why plant this grass? "The marsh grass will stop some of the erosion and filter out toxic materials that come off the land," explained Dr. Neidhardt.

The students also built wood duck nesting boxes in the marsh, took a canoe trip around the island, and met with local wildlife workers and foresters from the Maryland Department of Natural Resources. They even had an evening of entertainment provided by a well-known Chesapeake folksinger.

For the long hours of work and planning that the students spent, they each received one-half credit toward graduation. They also got the satisfaction of seeing wildlife and improving part of the Chesapeake ecosystem. Shannon Bryant, for example, talked about what she and her classmates left behind. "I'll be able to picnic here one day and see our grass."

The students now agree that more people should get involved in working to conserve the environment. As Adam Halberstodt said, "I think we take a lot from the earth without even thinking about it. Working out there was the least I could do."

Glossary

abiotic nonliving parts of the environment that include chemicals in the air, water, and soil, 7

Antarctic the southern polar region of the earth, 41

Arctic the northern polar region of the earth, 41

autotrophs organisms that make carbon nutrients by absorbing carbon dioxide from the environment, 11

biomagnification an increase in the amount of harmful chemicals absorbed by each organism moving up the food chain from producer to top consumer, 28

biotic pertaining to living things, 11

carnivores animals that eat only other animals, 26

chemosynthesis a process in which organisms make carbohydrates using energy obtained by combining oxygen with such inorganic chemicals as ammonia, sulfur, and iron oxides, 25

climax community a community that keeps a fairly constant number and variety of organisms for a long period of time, 63

competition the struggle of different species to inhabit the same space and use the same food supply and the struggle between individuals of the same species to use the same resources in the environment, 34

composting the breakdown of organic materials in solid waste to form a mixture of decayed leaves and fertilizer used for improving soil, 48

condensation the physical change of a gas or vapor to a liquid, 41

consumers organisms that gain energy by eating other organisms, 26

culture the arts and customs that make up a way of life for a group of people at a certain time, 15

decomposers bacteria and other consumer organisms that break down matter into simpler parts, 26

disperse to move, break up, or spread in the environment, 57

domesticate to change animals or plants so that they can live with or be used by humans, 15

ecological succession a change in the number and species of living things in a specific ecosystem, brought about as certain plant and animal species gradually take over an area from other species, 62

evaporation physical change from a liquid into a gas, 41

food chain the order in which living things in a community of organisms feed upon one another; a sequence of feeding interactions in which consumer organisms feed upon producers and other consumers, 26

food web the connection of food chains within a community of organisms, 26

glacier a large moving mass of ice formed by snow that does not melt. It travels slowly along a valley and either advances or retreats depending upon the amount of snow falling at its source, 41

groundwater water below the earth's surface that supplies springs and wells, 42

herbivores animals that eat only plants, 26

heterotrophs organisms that cannot make their own food, but are dependent on other organic matter for food, 11

legumes plants that make a seed in a pod, such as bean plants, 47

limiting factors conditions of the environment that limit the growth of a species, 36

microbes very tiny organisms that can be seen only through a microscope, 6

nitrate one of several chemical compounds made from the elements nitrogen and oxygen, 47

nutrients substances from the environment used by organisms for life and growth, 7

omnivores animals that eat both plants and animals, 26

organic any chemical compound or matter containing carbon, 48

photosynthesis a chemical process during which light energy is absorbed by a pigment in plants and is used for making carbohydrates from carbon dioxide and water, 24

phytoplankton tiny plant life found floating in the ocean or in bodies of fresh water, 25

pioneer species the first kind of plant or animal to move into and inhabit an area barren of life forms, 62

precipitation the falling of water in the form of rain, sleet, hail, or snow, 41

predation the process in which one animal hunts another animal for food, 32

species diversity a measurement that includes the number of different organisms living in one area and the number of individuals making up each species, 63

technology the use of tools, machinery, and materials to produce the goods and services to satisy human needs, 15

transpiration the transfer of water, as a gas, from plant leaves to the atmosphere, 42

trophic level describes the position of the organism in relation to the solar energy that flows through an ecosystem, 27

zooplankton small animals found drifting in oceans or bodies of fresh water, 26

Resource Directory

Academy of Natural Sciences
19th and the Parkway, Legion Square
Philadelphia, PA 19103
The oldest museum of natural history in North America.

American Forests
P.O. Box 2000
Washington, D.C. 20013
A special research and education program of the American Forestry Association that sponsors research on the ecology of forest ecosystems; supervises Global Releaf programs through which people and corporations help reforest damaged woodland in urban and rural areas.

American Museum of Natural History
Central Park West at 79th St.
New York, NY 10024
Features realistic exhibits that show wildlife in all natural ecosystems.

American Wildlands
3609 S. Wadsworth Blvd.
Lakewood, CO 80235
Supports the protection of wetland, river, and forest ecosystems in the United States; conducts ecological research and publishes free materials for students.

Audubon Naturalist Society of the Central Atlantic States
8940 Jones Mill Rd.
Chevy Chase, MD 20815
Supports the protection of wildlife and natural ecosystems in the Mid-Atlantic states; conducts public field trips and sponsors an outdoor natural history program for schools in Washington, D.C.

Audubon Society of New Hampshire
3 Silk Farm Rd.
P. O. Box 528 B
Concord NH 03302-0516

James Ford Bell Museum of Natural History
17th and University Ave., SE
Minneapolis, MN 55455
Features dioramas of the different ecosystems, as well as the plants and animals of Minnesota; includes the northern evergreen forest, prairie, and southern deciduous forest.

Belle Isle Nature Center
Belle Isle Park
Detroit MI 48207
Conducts environmental programs in a 300-acre forest and wetland ecosystem.

Center for Marine Conservation
1725 DeSales St., NW
Washington, D.C. 20036
Supports the protection of endangered marine wildlife, such as whales and sea turtles; promotes the conservation of natural habitats in which endangered species live.

Chesapeake Bay Foundation
162 Prince George St.
Annapolis MD 21401
Provides education about the Chesapeake Bay ecosystem.

Coast Alliance
235 Pennsylvania Ave., SE
Washington, D.C. 20003
Informs the public about the value of coastal ecosystems; protects and conserves wildlife and their habitats by supporting government laws and programs.

Cousteau Society
777 Third Ave.
New York, NY 10017
Explores and conducts research on the world's oceans; publishes books and makes films that support the protection of the ocean ecosystem.

Deep Portage Conservation Reserve
Route 1, Box 129
Hackensack MN 56452

Operates a 6000-acre forest ecosystem and environmental education campus with classrooms, museum, trails, and theater.

Desert Fishes Council
P.O. Box 337
Bishop, CA 93515

Originally concerned with preserving the native fish of the Death Valley region of California. Currently trying to protect desert ecosystems in the southwestern United States and northern Mexico.

Florida Audubon Society
460 Highway 436
Suite 200
Casselberry, FL 32707

Offers educational programs about Florida ecosystems.

Friends of the Everglades
101 Westward Dr., No.2
Miami Springs, FL 33166

Promotes the preservation of the Florida Everglades and Upper Key Largo, which contains the only living coral reef ecosystem in the United States.

Greater Ecosystem Alliance
P. O. Box 2813
Bellingham WA 98227

Promotes conservation of ecosystems in northwestern United States.

Greater Yellowstone Coalition
13 South Willson
Bozeman MT 59771

Works to raise public awareness of ecosystems, especially the Yellowstone ecosystem.

High Desert Museum
59800 S. Highway 97
Bend OR 97702

Focuses on high desert ecosystems of eight western states.

John Inskeep Environmental Learning Center
19600 S. Molalla Ave.
Oregon City, OR 97045
(503) 656-0155

Promotes understanding of ecosystems for people of northwestern United States; provides environmental education experiences, resources, and materials.

Institute of Ecosystem Studies
Cary Arboretum
Millbrook NY 12545-0129

Offers educational programs and publications.

Kentucky Association for Environmental Education
Blackacre Nature Preserve
3200 Tucker Station Rd.
Jeffersontown, KY 40299

Provides environmental programs throughout Kentucky.

Maine Audubon Society
Gilsland Farm, P. O. Box 6009
Falmouth ME 04105

Operates environmental education programs, library, and resource center. Provides information about forest and marine ecosystems.

Massachusetts Audubon Society
S. Great Rd.
Lincoln MA 01773

Protects 18 sanctuaries and offers environmental education.

Michigan Audubon Society
6011 W. St. Joseph, suite 403
P. O. Box 80527
Lansing MI 48908-0527

Works to protect Great Lakes ecosystems and offers environmental education through five major centers.

National Audubon Society
666 Pennsylvania Ave., SE
Washington, D.C. 20003-4319

Provides educational programs and publications, and conducts ecosystems studies throughout the United States.

National Institute for Urban Wildlife
10921 Trotting Ridge Way
Columbia, MD 21044
Conducts research on the relationship between human beings and wildlife in urban ecosystems; publishes educational manuals about urban wildlife for students and teachers.

National Wildlife Federation
1412 16th St., NW
Washington, D.C. 20036
Conducts research to investigate the status of endangered species and other wildlife; publishes books and sponsors educational programs.

New Jersey Audubon Society
790 Ewing Ave.
Franklin Lakes, NJ 07417
Offers educational programs and operates sanctuaries.

Northcoast Environmental Center
879 Ninth Street
Arcata, CA 95521
Provides environmental information about northern California, operates a library open to the public, and conducts special projects on forest ecology.

Pocono Environmental Education Center
R.D. 2, P.O. Box 1010
Dingman's Ferry, PA 18328
(717) 828-2319
Provides on-site environmental programs for visitors to their 38 acre campus, and access to about 200,000 acres of adjacent public parkland.

Roger Tory Peterson Institute of Natural History
110 Marvin Parkway
Jamestown, NY 14701
Provides educational programs and publishes books, magazines, and newsletters about wildlife, wildlife habitats, and natural history.

Student Conservation Association
Box 550, Charlestown, NH 03603
Recruits high-school youth for volunteer field work and training in conservation; cooperates with public and private land/resource agencies.

Urban Habitat Program
300 Broadway, #28
San Francisco, CA 94133
Develops and publishes educational materials about the environment in the San Francisco bay area.

Virginia Museum of Natural History
1001 Douglas Ave.
Martinsville, VA 24112
Offers educational programs on Virginia's natural heritage.

The Wildlife Conservation Society
2300 Southern Blvd.
Bronx, NY 10460
The society, which operates the New York Aquarium and Bronx Zoo, offers innovative educational programs for schools and the general public.

World Wildlife Fund
1601 Connecticut Ave., NW
Washington, D.C. 20037
Provides programs and services to promote the protection of endangered species of plants and animals; purchases natural areas to preserve particular species.

Index

G

H

I - K

L

M

N

O

P

R

S

T

U

V

W

Y

Z

Photo Acknowledgments

2: Randall E. Raymond, UEEID, Cass Technical High School; **3:** Randall E. Raymond, UEEID, Cass Technical High School; **5:** Macdonald Photography, Picture Cube; **6:** © Jeffrey High, Image Productions; **16:** United Nations; **17:** United Nations; **20:** David Manski; **21:** David Manski; **28:** Arthur C. Smith III, Grant Heilman; **33:** Grant Heilman; **34:** © Gordon S. Smith from the National Audubon Society, Photo Researchers; **35:** Hal Harrison, Grant Heilman; **36:** Ralph H. Anderson, National Park Service; **37:** © Lemand Lee Rue from the National Audubon Society, Photo Researchers; **40:** Grant Heilman; **43:** ©NYZS/The Wildlife Conservation Society; **45:** Grant Heilman; **45:** © Bernard Pierre Wolff, Photo Researchers; **53:** Ann Rodman, National Park Service; **54:** Grant Heilman; **56:** Jim Peace, National Park Service; **57:** Grant Heilman; **58:** Grant Heilman; **59:** Grant Heilman; **62:** © Woodrow Goodpaster from the National Audubon Society, Photo Researchers; **68:** Grant Heilman; **71:** Douglas Falk; **72:** Grant Heilman; **73:** Grant Heilman; **74:** Douglas Falk; **75:** Florida State Archive; **81:** Lynne Carnes.